Moses Quinby

Mysteries of Bee-Keeping Explained

Containing the Result of Thirty-Five years' Experience, and Directions for Using the

Movable Comb and Box-Hive

Moses Quinby

Mysteries of Bee-Keeping Explained

Containing the Result of Thirty-Five years' Experience, and Directions for Using the Movable Comb and Box-Hive

ISBN/EAN: 9783337228477

Printed in Europe, USA, Canada, Australia, Japan

Cover: Foto ©Lupo / pixelio.de

More available books at **www.hansebooks.com**

NEWLY WRITTEN THROUGHOUT.

MYSTERIES
OF
BEE-KEEPING
EXPLAINED.

CONTAINING THE RESULT

OF

THIRTY-FIVE YEARS' EXPERIENCE, AND DIRECTIONS FOR USING THE MOVABLE COMB AND BOX-HIVE, TOGETHER WITH THE MOST APPROVED METHODS OF PROPAGATING THE ITALIAN BEE.

BY

M. QUINBY,

PRACTICAL BEE-KEEPER.

NEW STEREOTYPED AND ILLUSTRATED EDITION.

NEW-YORK:
ORANGE JUDD & COMPANY.
1866.

Entered according to Act of Congress, in the year 1865, by
M. QUIMBY,
in the Clerk's Office of the District Court of the United States for the Southern District of New-York.

CONTENTS.

CHAPTER I.
INTRODUCTORY REMARKS—17

Description of queen.................. 17
Age of queen 18
Workers...................................... 19
Age of worker............................ 19
Drones.. 19
Age of drone.............................. 20
Preparations for swarming........ 20
Their nature should be understood.... 20

CHAPTER II.
PHYSIOLOGY AND BREEDING.—22

Imperfectly understood............... 22
When they begin to rear brood... 22
How small colonies begin........... 23
Different in large families........... 23
Laying 24
When the eggs hatch.................. 25
Rough treatment 25
Time before the young bee commences labor............... 26
Terms applied to young bees..... 27
Time from the egg to mature queen.. 27
Short cells usually taken for rearing queens....................... 29
When the queen leaves to meet the drone........................... 29
Number of eggs that a queen will lay. 30
When drones are reared............. 33
When queens are reared in swarming hives........................ 33
When queens and drones are destroyed before swarming........ 34
Queen leaves with the first swarm.. 35
What becomes of the bees when no swarm issues........................ 35
When a young queen takes the place of the old one........................ 35
When large numbers of drones are reared................................... 35
Theory relative to drones........... 36
Mr. Wagner's Theory.................. 39
Mr. Harbison's Theory............... 40

CHAPTER III.
HIVES.—46

Satisfaction in having no patent....... 46
No difficulty in obtaining certificates and premiums 47
Hives furnished for trial.............. 48
Necessities of the bees................ 48
Great discovery for patent venders... 49
Principles of different hives....... 49
Chamber hive............................. 49
Suspended hive.......................... 50
Inclined bottom board............... 50
Dividing hive............................. 50
Changeable hive........................ 51
Ventilating hive......................... 54
Moth proof hive........................ 54
Non-Swarmer............................ 55
Common box hive..................... 58

Proper size of hive.................... 59
Directions for making box hive......... 61
Top of hive not fastened............... 62
Best surplus boxes. 63
Directions for making boxes............ 63
Guide comb............................. 64
Some desirable things not found in box hive................................ 65
Movable comb hive...................... 66
Some of its advantages................. 67
Movable comb hive as used by the author................................. 68
Directions for making.................. 63
Straw hive for wintering bees.......... 73
Observatory hive....................... 75

CHAPTER IV.

BEE PASTURAGE.—76

Substitute for pollen.................. 76
Substitute for honey................... 77
Manner of packing pollen............... 77
Flowers that yield first pollen........ 78
First honey............................ 79
Fruit flowers important................ 79
Red raspberry a favorite............... 80
Honey from red clover.................. 80
Catnip one of the best honey yielding plants............................... 80
Singular fatality attendant on Silkweed............................... 82
Bass-wood very important............... 83
Honey-dew.............................. 85
Unusual secretion 86
Buckwheat.............................. 89
Do bees injure the grain?.............. 89
Bees necessary to insure a crop........ 90
Two kinds of pollen stored in one cell 92
No test of the presence of a queen..... 93
Bee-bread seldom packed in drone cells 94
Manner of discharging pollen........... 94
Discharging honey...................... 94
Some cells contain honey for daily use 95
Combs constructed as needed............ 96
Best season for honey.................. 96
How many stocks may be kept............ 97
Principal sources of honey............. 99
Distance a bee will go for honey.......100

CHAPTER V.

THE APIARY.—100

Location...............................100
Location marked........................102
Should not be moved....................102
Space between stands...................103
Small matters..........................103
Cheap stand............................104
Disadvantage of standing too high......105
Best cover.............................107
Bee-house unprofitable.................107
Some will have them....................107
Hives should be of different colors....107
Replacing queens.......................109
Several bee-houses.....................111

CHAPTER VI.

ROBBING.—113

Not understood 113
Difficulty in deciding.................115
Weak colonies in danger................115
When to look out for robbers...........116
First indications......................117
Remedies...............................118
Equalization...........................120
Battles................................121

CHAPTER VII.

FEEDING.—122

Feeding a last resort..................122
Care...................................123
Destitute colonies sometimes desert...124
When they must be fed..................124
Manner of feeding......................126
Object in feeding......................127
Promiscuous feeding unprofitable......129

CONTENTS.   v

CHAPTER VIII.
DESTRUCTION OF THE MOTH WORM.—129

Some in best stocks................129
Fear of the bee....................130
How destroyed.....................131
Moth proof hive not made..........134
Box for wren......................135

CHAPTER IX.
PUTTING ON AND TAKING OFF BOXES.—135

Must not be put on too early.........135
Making holes after the hive is full ...137
Boxes may be too easy of access 138
A better way........................139
Advantage of glass-boxes............139
When to take off....................140
How to get rid of the bees..........141
Bees not disposed to sting..........141
To secure honey from worms..........143
The way the worms get in............144
Remedy..............................145

CHAPTER X.
SWARMING.—146

Knowledge necessary................146
When swarming commences............147
Indications.......................148
Care in examining hives...........149
Preparations for swarming.........150
When swarms issue.................151
Why drones are sometimes killed in spring.........................150
Which bees issue..................153
The old queen leaves..............153
Hives should be ready.............154
Immediate indications of a swarm..154
Swarm clusters...................155
How to do it.....................155
All should be made to enter......159
Carry to the stand...............160
Put in movable comb hive.........160
Shade important..................160
Clustering bushes................161
Loss by flight...................162
Nothing but bees necessary in a hive..163
Do they select a home before swarming?..........................164
How far will they go?............164
One first swarm has bees enough..165
How to keep separate.............166
Cannot be stopped when parts are on the way........................167
How to divide....................167
Different process with movable comb-hive...........................171
Swarms sometimes return..........173
First swarms choose good weather..174
Exceptions.......................175
After-swarms.....................175
Their size.......................175
When expected....................175
Piping of the Queen..............176
Variation in time of issuing.....177
How after-swarms issue...........178
Number of queens.................179
Do not always choose good weather..179
Go farther before alighting......179
Propriety of returning...........180
Moth worm troubles small colonies..181
Uniting..........................182
More trouble.....................182
Rule.............................183
One queen destroys others........184

CHAPTER XI.
ARTIFICIAL SWARMS.—185

Perplexities.....................185
Work well........................186
Do it in season..................186
First experience.................186
How to make artificial swarms....188
Manner of placing the stands.....189
Queen-cell, to introduce.........190
Operations with movable combs easy..191
One division will make drone comb..191
Too many drone combs for profit..192
Honey made in boxes in the hive..192
Boxes transferred and finished on another hive.......................193
Time for queen to lay............193

CHAPTER XII.

LOSS OF QUEENS.—195

When lost by swarms....................195	Result..200
Drone-comb..................................195	Age of bees...................................201
Speculations.................................195	Duty...201
Disputed question.........................196	Remedy. 201
A multitude of drones needed........197	Mark date of swarm......................203
When the loss occurs....................198	Other remedies............................203
Time of leaving varies...................199	Indications of loss in early spring....203
Indication of loss..........................200	

CHAPTER XIII.

PRUNING.—205

Seldom necessary..........................205	Best time......................................207
The time..205	Little risk of stings.........................209
Difficulty in driving in cool weather...207	Frequent pruning not recommended..209

CHAPTER XIV.

DISEASED BROOD.—210

What is it?....................................210	Supposed cause............................212
Italians less affected 210	How it spreads..............................216
Where found.................................211	Theory..217
When first discovered...................211	Caution...218
Description...................................211	Examination..................................218
Remedies attempted......................212	Assumed knowledge 220

CHAPTER XV.

ANGER OF BEES.—221

Causes of irritability 221	Bee-charms unreliable..................225
How they make an attack..............223	Sting..226
Never irritable when after honey....223	Does its loss prove fatal ?.............226
Smoker described.........................224	Protection....................................227
Italians less docile........................225	Remedies for stings......................227

CHAPTER XVI.

ENEMIES OF BEES.—229

Rats and mice................................229	Worms sometimes work in centre of comb..237
A word for King-bird......................229	Bees mutilated by webs.................237
Chickens will eat drones................230	Bees fastened in the cells.............238
Cat-bird acquitted.........................230	Different appearance in old stocks...239
Toad...231	Worms grow larger when undisturbed..240
Black wasp...................................232	Freezing destroys them.................241
Ants—a word in their favor...........232	Extermination of the moth.............241
Spiders..234	Seldom exempt in ordinary management 243
Moth...234	Remedies.....................................244
Where their eggs are deposited.....236	

CHAPTER XVII.
WAX.—245

What is it?.... 246
How it is obtained....................246
Commencement of a comb............247
Crooked combs.........................250
Straight combs.........................250
Quantity of honey taken by a swarm..250
Making drone-cells....................251
Some wax wasted... 251
Water necessary........................252
Cells uniform in size..................252
Melting of combs.......................254

CHAPTER XVIII.
PROPOLIS.—256

How obtained...........................256
How discharged........................256
New swarms sometimes use wax instead................................... 257
More abundant in August..............258

CHAPTER XIX.
TRANSFERRING.—259

Preparation............................259
Time when..............................259
How to do it...........................260
Keep brood together....................260
Caution................................262

CHAPTER XX.
SAGACITY OF BEES.—263

Too marvellous.........................263
Instances of sagacity..................264
No part of the hive inaccessible......265
We should be content with facts.......266

CHAPTER XXI.
SELECTING COLONIES FOR WINTER.—266

First care............................266
Strong colonies inclined to rob........267
Requisites of good stocks.............267
Disadvantage to kill bees..............268
Cause of poor colonies varies in different sections........................268
Poor stocks may be united............269
When it is not best....................269
Two swarms united, eat less then when separate..............................270
Season to operate.....................271
Paralyzing bees.......................271
Description of fumigator..............271
How to operate.........................272
How bees were wintered in a scarcity of honey.............................274
Advantage of transferring.............274
Uniting comb, honey, and bees........275
When it is best to feed................277

CHAPTER XXII.
STRAINING HONEY AND WAX.—279

Removing combs.........................279
How to strain 279
Metheglin and vinegar.................281
Feeding refuse honey..................281
Making wax............................282
Quantity wasted 282
Large quantities......................285

CHAPTER XXIII.
WINTERING BEES.—284

Different methods.....................284
Warmth requisite......................284
Size of colony........................285
Setting out............................297
A building for the purpose............298
Room in dwelling house 298

Promotion of warmth.................285
Moisture............................286
Causes of starving..................287
Dysentery...........................287
Water...............................292
Natives of a warm climate...........294
Warm room...........................294
Cellar preferred....................295
Housing.............................296
Burying bees........................299
Straw hives.........................300
Philosophy..........................300
Straw top...........................300
Simple box..........................304
Mice................................304
Shade...............................305
Lost on snow........................305

CHAPTER XXIV.
THE ITALIAN OR LIGURIAN BEE.—308

Reputation..........................308
Importers...........................308
Superiority.........................308
When first obtained.................309
Object..............................310
Peculiarities.......................311
Longer lived........................312
Robbing.............................313
Disposition.........................314
Swarming............................315
Hive crowded with bees in cool weather.........................317
Remedy..............................317
Purity to be secured................318
Variation in color of queens........319
Susceptible of improvement..........319
Neighbors join in purchasing queens.321
Mix three miles distant.............321
Colony to furnish drones............321
Method of Italianizing a whole apiary.322
Artificial queens...................323
How to rear them....................323
How to obtain bees for rearing queens.325
Black bees as nurses................326
Best time to obtain brood...........327
Finding queen.......................328
Introduction of queen...............329
Italianizing the box-hive...........330
Test of the presence of the queen...331
Transporting queen..................332

CHAPTER XXV.
PURCHASING STOCKS AND TRANSPORTING BEES.—333

Qualification for an apiarian.......333
Luck................................333
Purchase the best...................336
Avoid diseased stocks...............336
Old ones not objectionable..........336
Transporting bees...................338

CHAPTER XXVI.
CONCLUSION.—340

PREFACE.

One who for thirty-five consecutive years has succeeded in keeping bees, and has been able, most of that time, to count his stocks by hundreds, can hardly fail to furnish something from his experience, that will be beneficial and interesting to others; and he will doubtless be pardoned for attempting to teach those who may desire to avail themselves of his knowledge, and thus avoid the tedious process of acquiring it for themselves.

Twelve years ago the author explained some of the "Mysteries of Bee-keeping," to the public. The simple, practical and comprehensible instructions given, have met with abundant favor among old practical bee-keepers, and interested thousands who are now keeping bees with decided success.

The greater number of bees kept, the increased quantity and improved appearance of the honey in our markets, encourage the belief that many who have at present no adequate conception of the immense annual waste of this delicious production, may yet be induced to make an effort to save a still greater proportion of it. It will not be pretended that such immense numbers of bees may be kept in any particular section of the United States, as are reported to prosper in some parts of Europe, (2000 hives to the square mile,) but no one will deny that hundreds of thousands of stock might be profitably added to the present amount. It has been estimated that on an average, every acre will produce its pound of honey. New York alone contains 30,000,000 acres. Shall we suffer this enormous loss of the gifts of a beneficient Creator, without an effort to secure to ourselves and the community, so valuable and vast a treasure? All that is necessary, is sufficient encouragement and knowledge of the subject.

Enough has already been done to show that the estimate is sufficiently near the truth to be taken as a base for future calculations. An area of a few square miles in the writer's vicinity, has, in some favorable seasons, furnished for market over 20,000 lbs. surplus honey. Had a proportional quantity been collected in all other places in the United States, we could count the proceeds by millions of dollars instead of a few hundreds or thousands.

The author does not offer this improved edition because he supposes that people would be unable to keep bees without it, but with the hope that those who are already doing well, may do better. A person who wishes to make the most possible from his bees can hardly afford to dispense with the benefit of any experience that will aid him. The instructions found in the periodicals of the day are often not to be depended upon. A score of bee-keepers, each of limited experience, will give as many different methods, and an editor equally inexperienced, is usually unable to discriminate between them. The simplest directions of a reliable practical bee-keeper who studies the science with an honest enthusiasm, are invaluable to the tyro in apiarian knowledge.

To benefit the largest possible class, the author has endeavored to be practical rather than scientific, and has aimed at no elegance of style or diction, preferring that the merit of the book should lie in its simplicity and reliability. M. QUINBY,

St. Johnsville, N. Y., April, 1865.

ILLUSTRATIONS.

Queen Bee	17
Regard of the Worker for the Queen	18
Worker Bee	19
Drone Bee	20
Brood from a Drone Queen in Worker Cells	37
Comb showing the different shape of Cells, when an attempt is made to raise Queens from Drone Brood	38
Ovaries of the Queen	41
Roof	65
Simple movable Comb Hive	69
Movable Frame	70
Wire braces to support Frames	70
Honey Board	71
Straw Hive for Winter	74
Bee House	108
Bee House	110
Bee House	111
Hives arranged in a hedge	113
Feeder	126
Worm Gallery in the Comb	130
Moth-worm	130—240
Tool for killing Worms	131
Cluster of Queen Cells	150
Bag for Hiving Bees	158
Frames to hold Boxes inside the Hive	192
Tools for Pruning	208
Bee Hat	227
Bee Moth	235
Worm Gallery removed from the Comb	237
Cocoons of the Moth Worm	240
Abdomen of the Bee, enlarged, showing the scales of Wax	247
Piece used to steady the Combs in Transportation and Transferring	259
Transferred Combs	261
Fumigator	272
Comb containing Brood from which to raise Queens	323
No. 34 Inserted in a Comb ready for the rearing box	324
Queen Cells made on such Comb	325

CHAPTER I.

INTRODUCTORY REMARKS.

Every prosperous swarm or family of bees must contain one queen, several thousand workers, and part of the year, a few hundred drones.

DESCRIPTION OF QUEEN.

The *Queen* is the mother of the entire colony. Her only duty seems to be to lay eggs, of which she sometimes deposits two thousand in twenty-four hours. In shape, she resembles the worker more than the drone, but is longer than either, and like the worker possesses a sting, but will not use it upon anything below royalty. Her color upon the upper side is darker than that of the others;

Fig. 1.—QUEEN.

the two posterior legs and under side are of a bright copper color. In some of them a yellow stripe nearly encircles the abdomen at the joints. All the colors are bright and glossy, and she has but little of the down or hair that is seen on the drones and workers. Different queens vary much in color, some being much darker than others. A still greater variation is presented in the Italian queens, most of which are of a rich golden color, while a few are even darker than the usual shade of the natives.

For the first few days after leaving the cell her size is much less than after she has assumed her maternal duties. She seldom, perhaps never, leaves the hive, except when leading out a swarm, and when but a few days old to meet the drone for the purpose of fecundation.

18   INTRODUCTORY REMARKS.

AGE OF QUEENS—THEIR OFFICE.

The average age attained by the queen is about three years. The idea that the queen governs the colony, and directs all their operations, is probably totally erroneous. They manifest a certain regard and affection for her, how-

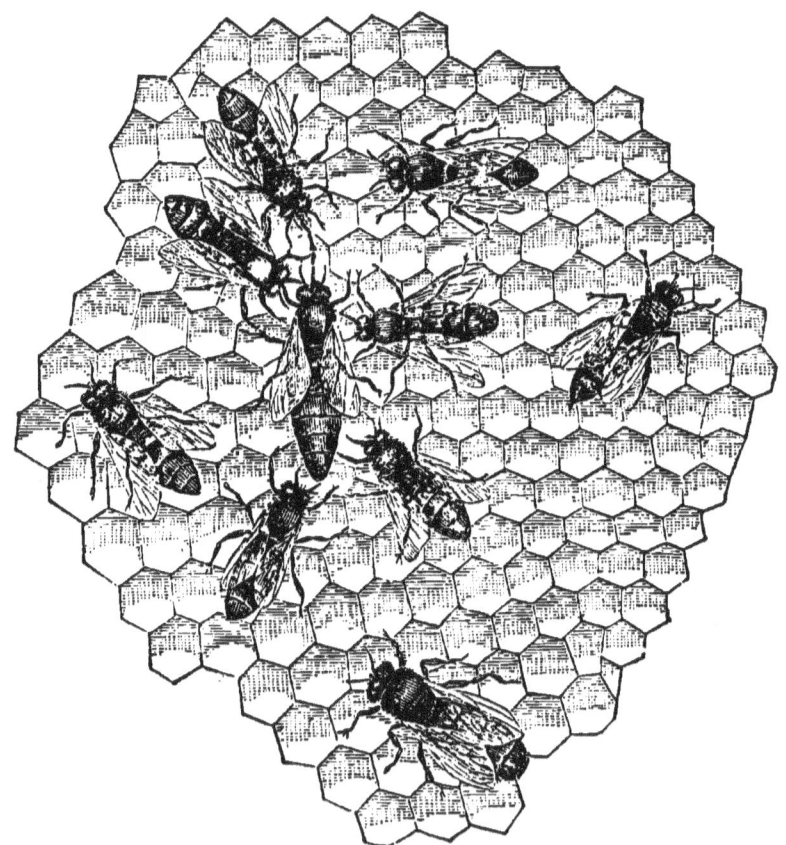

Fig. 2.—REGARD OF THE WORKERS FOR THE QUEEN.

ever, and a half dozen or so may often be seen gathered around her, as shown in the accompanying cut. They will, when destitute of a queen, continue their labors with as much system and regularity, as when one is present, although she is necessary to insure their permanent prosperity.

WORKERS.

All labor devolves on the *Workers*. These are provided with a sac or bag for gathering honey, and basket-like cavities on their posterior legs in which to pack the pollen  of the flowers in little pellets for carrying it home to the hive. They range the fields for honey and pollen, secrete wax, construct combs, prepare food to nurse the young, bring water, obtain propolis to seal up all

Fig.3. WORKER. crevices and flaws about the hive, stand guard to keep out intruders, etc.

Huber and some others divide the workers into classes, such as wax-workers, pollen-gatherers, nurses, etc., but it is very difficult to believe that any such distinctions exist.

For the defence of their treasures and themselves, they are provided with a sting and a virulent poison, but will not use it when abroad, if unmolested; they volunteer an attack only when near the hive. They are all females with undeveloped organs of generation, yet they possess enough of the maternal instinct to make them good nurses for the brood of the real mother. For nearly two weeks after the young worker emerges from its cell, it is almost exclusively engaged within the hive; thereafter, it assists in collecting stores.

AGE OF THE WORKER.

Its age varies from one to eight months, according to the season in which it is hatched. In the busiest season it lives but a few weeks, but when hatched at the beginning of cool weather, its life is extended several months.

DRONES.

The *Drones* are the males; their bodies are large and clumsy, and without the symmetry of the queen and worker. Their buzzing when on the wing is loud, and different from that of the workers. They have no sting, and may be taken in the fingers with impunity. They seem to be of the least valuable class in the bee community; they assist sometimes, in keeping up the necessary animal heat in the hive; but one only, out of thousands, is actually serviceable in fecundating the queen. The number reared depends upon the strength of the colony, and the stores on hand or being collected.

Fig. 4.—DRONE.

AGE OF DRONES.

Whenever a scarcity of honey occurs, they are all destroyed. Thus their life is very precarious, being sometimes limited to a few hours, or extended to a few days, weeks or months; but averaging much less than that of the workers.

PREPARATIONS FOR SWARMING.

In the spring and early part of summer, when nearly all the combs are empty, and food is abundant, the bees rear brood more extensively than at any other period. The hive soon becomes crowded with bees, and royal cells are constructed, in which to raise queens. When some of these young queens are sufficiently advanced to be sealed over, the old one, and the greater part of the workers, leave for a new location, (termed swarming,) leaving those remaining to maintain the prosperity of the hive. They soon collect in a cluster, and if put into an empty hive, commence anew their labors, constructing combs, rearing brood, and storing honey, to be abandoned the following year, as before.

THE NATURE OF BEES SHOULD BE UNDERSTOOD.

We should fully understand that the nature of the bee, found under any circumstances, climate, or condition, is the same. Instincts first implanted by the Creator, have come unimpaired through millions of generations to the present day, and will continue unchanged.

To gratify our acquisitiveness, we have forced them to labor under every disadvantage; yes, we have compelled them to sacrifice their industry, their prosperity, and even their lives, but they have never yielded their instincts. We may destroy life, but cannot improve or change their nature; the laws that govern them are fixed and immutable.

Spring returns to its annual task, dissolves the frost and warms into life nature's dormant powers. Flowers, with a smile of joy expand their delicate petals in grateful thanks, while the stamens sustain upon their tapering points the anthers covered with the fertilizing pollen, and the pistil springs from a cup of liquid nectar, and the delicious fragrance imparted to every breeze, invites the bee as with a thousand tongues to the sumptuous banquet. She does not need any stimulus from man as an inducement to partake of the feast; without his aid she visits each cup of wasting sweetness, and secures the tiny drop, while the superabundant farina, dislodged from the nodding anthers, covers her body to be brushed together and kneaded into bread. All she requires at the hand of man, is a suitable storehouse for her treasures.

Industry is a part of bee nature. If, when their tenement is supplied with all things necessary to take them safely through the winter, and there is no necessity for continued labor, we furnish them additional room, they assiduously toil to fill it up. Rather than to pass their time in idleness, during a bounteous yield of honey, they will sometimes deposit their surplus in combs outside of the hive, or under the stand. This inherent industry lies at the foundation of all the advantages in bee-keeping, consequently our hives must be constructed with this end in view; but at the same time we must not interfere with other requirements of their nature. Their peculiar traits mentioned in this chapter, will be more fully discussed in different parts of this work, as appears to be called for, where proof will be offered to sustain the positions here assumed, which are as yet mere assertions.

CHAPTER II.

PHYSIOLOGY AND BREEDING.

IMPERFECTLY UNDERSTOOD.

Comparatively few people have a very definite idea of the time and manner of rearing brood. Many persons who have kept bees for years have bestowed so little attention upon this point that they are unable to tell at what time they commence, how they progress, or when they cease. They have an idea that one swarm, and occasionally two or three, is reared sometime in June, or the early part of summer, and this comprises their whole knowledge of the subject. Whether the drones deposit the eggs, or some of the workers are females, and each raises one or two, or the "king bee" is a common parent of the eggs, is quite beyond their ability to decide. It is hardly necessary to inform observing apiarians that the queen is the mother of the whole family.

WHEN THEY BEGIN TO REAR BROOD.

The period at which she commences depositing eggs probably depends on the strength of the colony, amount of honey on hand, etc., and not upon the time when gathering food begins. Strong colonies frequently begin to rear brood by Christmas. When sweeping out the litter under the hives as early as the first of March, young bees may often be found under the best stocks. Observation shows that there is but little time when our best colonies have no brood—seldom more than two months. Yet stocks when very weak do not commence until warm weather. It seems that a certain degree of warmth is necessary to perfect the brood, and this a small family can not generate.

HOW BREEDING IS DONE IN SMALL COLONIES.

In a small family, the first eggs are deposited in the centre of the cluster of bees; it may not be in the centre of the hive in all cases, but the middle of the cluster wherever it chances to be located, is the warmest place. Here the queen will commence; a space not larger than a dollar is first used, and the cells exactly opposite on the same comb are next occupied. If there is sufficient warmth in the hive, produced either by warm weather or generated by the bees, she will then fill a spot on the adjoining combs corresponding with the first, but not quite as large. The circle of eggs in the first comb is then enlarged, and more added to the next, and so on, continuing to spread, and keeping the distance from the center to the outside of the space occupied by eggs about equal on all sides, until they occupy the entire surface of the comb. Long before the outer edge is occupied, the first eggs deposited are matured, and the queen returns to the centre and uses these cells again, but she is not as particular this time to fill so many in exact order as at first, though with the Italian queen the brood is always very compact. This is the general process with small families. I have removed the bees from such in all stages of breeding, and have always found their proceedings as described.

THE PROCESS DIFFERENT IN LARGE FAMILIES.

In very large families their proceedings are different. As any part of the cluster of bees is warm enough for breeding, there is less necessity for economizing heat, and confining all the eggs to one small spot, and some unoccupied cells will be found among the brood, and a few will contain honey and bee-bread. But in the breeding season, a circle of cells, an inch or two wide, containing bee-bread, borders the sheets of comb containing brood. As bee-bread is probably the principal food of the young bee, it is thus very convenient.

LAYING.

When pollen is abundant, and the swarm is in prosperous condition, they soon reach the outside sheets of comb with the brood. At this period, when the hive is about full, and the queen is forced to the outside combs to find a place for her eggs, it is interesting to witness her operations in a glass hive. I have seen her several times in one day on a piece of comb next the glass. The light has no immediate effect upon her, as she will quietly continue about her duty, not the least embarrassed by curious eyes at the window. I have frequently lifted out a comb on which an Italian queen was engaged in laying, without interrupting her in the least. Before depositing an egg she enters the cell head first, probably to ascertain if it is in proper condition, as a cell part filled with bee-bread or honey is never used. When a cell is ready to receive the egg, on withdrawing her head, she immediately curves her abdomen, and inserts it. After a few seconds she leaves the cell, when an egg may be seen attached by one end to the bottom. It is about one-sixteenth of an inch in length, slightly curved, very small, nearly uniform the whole length, abruptly rounded at the ends, semi-transparent, and covered with a very thin and delicate coat, which will often break at the slightest touch.

WHEN THE EGGS HATCH.

After the egg has been in the cell about three days, a small white worm may be seen coiled in the bottom, surrounded by a milk-like substance, which, without doubt, is its food. How this food is prepared is mere conjecture. The supposition is that is chiefly composed of pollen; this is strongly indicated by the quantity which accumulates in hives that lose their queen and rear no brood—that is, when a requisite number of workers is left. The workers may be seen entering the cell every few minutes, probably to supply this food. When the comb in the glass hive is new and white, these operations can be seen more distinctly than when it is old and dark.

In about six days after the worm hatches, it is sealed over with a convex waxen lid. It is now hidden from our sight for about twelve days, when it bites off the cover, and comes forth a perfect bee. The period from the egg to the perfect bee varies from twenty to twenty-four days, averaging about twenty-one for workers, and twenty-four for drones. The temperature of the hive will vary somewhat with the atmosphere; it is also governed by the number of bees. A low temperature probably retards the development, while a high one facilitates it.

ROUGH TREATMENT OF THE YOUNG BEES.

There have been some amusing accounts of the assiduous attentions given to the young bee when it first emerges from the cell. It is said that "they lick it all over, feed it with honey," etc., as if wonderfully pleased with their acquisition. If any one expects to see any thing of this kind, he must watch a little closer than I have. I have seen hundreds when biting their way out, and instead of care or attention, they often receive rather rough treatment. The workers intent on other matters, will sometimes come in contact with one part way out of the cell, with force sufficient to almost dislocate its neck, yet they do not stop to see if any harm is done, or to beg pardon. The little sufferer, after this rude lesson, scrambles back as soon as possible out of the way, enlarges the prison door a little, and again attempts to emerge, with perhaps the same result; a dozen trials are often made before it succeeds. When it does leave, it seems like a stranger in a multitude, with no friend to counsel, or mother to direct. It wanders about uncared for and unheeded, and rarely finds one sufficiently benevolent to bestow even the necessaries of life. It is generally forced to learn the important lesson of looking out for itself, the day it leaves the cradle. A cell containing honey is sought for, where its immediate wants are all supplied.

TIME BEFORE THE YOUNG BEE COMMENCES LABOR.

Some have said that it would leave the hive for honey on the day it left the cell. Since the introduction of the Italian, we can determine this point very accurately by noting the day when the first one hatches, and also when the first one comes home loaded. It is seldom less than seven, and quite often fourteen days before they are thus seen. Some tell us, too, that after the bees seal over the cells containing the larvæ, " they immediately commence spinning their cocoons, which takes just about thirty-six hours." I think it very likely, but cannot imagine how it was determined. I do not possess optical acuteness to look into one of these cells after it is sealed over. Suppose we drive away the bees and open the cell to examine the interior: the little insect stops its labor in a moment, probably disturbed by the air and light. I never could detect one at work. Suppose we open these cells every hour after sealing, can we tell any thing about their progress by the appearance of these cocoons, or even tell when they are finished? The thickness of a dozen would not exceed that of common writing paper. It would be interesting to know how these particulars were ascertained, or whether they are simply surmises. When the bee leaves the cell, a cocoon remains, and that is about all we *know* concerning it.

TERMS APPLIED TO YOUNG BEES.

The young bee when it first leaves the egg, is termed a *grub, maggot, worm,* or *larva ;* from this state it changes to the shape of the perfect bee, which is said to be three days after finishing the cocoon. From the time of this change, till it is ready to leave the cell, the terms *nymph, pupa,* and *chrysalis,* are applied. The lid of the drone cell is rather more convex than that of the worker, and when removed by the young bee in working its way out, is left nearly perfect, being cut off around the edges; a coat or lining of silk keeps it whole. The covering of the worker cell is mostly wax, and is pretty well cut to pieces by the time the bee gets out. The covering to the queen cell is like that of the drone cell, but of greater diameter, and thicker, being lined with a little more silk.

TIME FROM THE EGG TO THE MATURE QUEEN.

The time in which an egg originally destined for a queen will mature, will not vary much from sixteen days; but when larvæ that are started as workers, are taken, there will often be a variation. All of the three kinds of bees remain in the egg form three or four days; then in the grub form for five or six, partaking of food, after which they are sealed up. When bees are deprived of a queen, and have means to rear another, they select such brood as will produce one in the shortest time. Give them a piece of comb containing eggs just laid, some two or three days old, larvæ just hatched, and some varying in age from one to five days, and the first cells made will probably be over the larvæ about four days old, and in ten days a queen will have matured. To insure the possession of a queen, they may afterwards begin several others, perhaps use some of the eggs, or some that were eggs when the comb was given them; but if the first queen hatches, she makes it her business to destroy all immature ones.

SIMILARITY OF QUEEN AND WORKER EGGS.

The *fact* that queens raised in this way seem to possess all the requisites of those raised in swarming hives, indicates that the eggs laid in worker and queen cells are all alike. It also gives rise to the idea with many modern writers, that all eggs for both queens and workers are laid in worker cells, and *transferred to queen cells when wanted there*. The antipathy of one queen towards another, although an immature one, and her own offspring, is thought sufficient to prevent her depositing eggs in these cells. Now, without sufficient evidence to be able to deny this positively, I must content myself with merely expressing a disbelief. I would like to say that *I do not believe that the bees ever remove an egg or larva from a worker to a queen cell.* For several years I have raised queens artificially by the hundred, in small queen boxes. In nearly *all the boxes*, there would be some queen cells in that stage of progress when it would be supposed that such transfer would be desirable. I have watched diligently and never yet discovered it. Whenever a queen has been raised, the egg or larva was in the cell when given to the bees, and the workers always changed or enlarged a common cell to a queen cell. The shape of the cell depends on the position of the comb from which it is made; if from a comb with cells of ordinary length, they are enlarged, lengthened, and turned downward. If the cells are not very deep, or are near the lower edge of the comb where there is abundant room to turn them down, the enlargement and change of direction will be made very soon after they decide upon making a queen of it.

SHORT CELLS USUALLY TAKEN FOR REARING QUEENS.

In swarming hives, whenever the bees decide on rearing queens, cells that are short like those on the lower edge of comb not completed, or on the side, seem to be preferred, and quite a number are often built close together.

Mr. Harbison has, I understand, patented the manner of introducing the piece of comb containing brood from which queens are to be reared. It is simply to place the comb with brood in a horizontal position, thus bringing the cell vertical, and save the bees the trouble of making a crooked cell. The young bee thus stands on its head like a young queen, during this period of its existence. He does not claim that this makes it a queen, but that from the same number of eggs, more queens will be raised. I have watched such combs with considerable interest, when side by side with a piece of comb placed the other way, with abundant room directly underside, without discovering that a larger number of queens was produced, or that they were any larger or better.

WHEN THE QUEEN LEAVES THE HIVE TO MEET THE DRONE.

In about six days after the queen has left her cell, if no competitors are in the way, she leaves the hive to meet the drone. I presume that it does not make much difference whether she has been reared in a large swarming hive, or in a small box particularly designed for rearing queens. The meeting takes place high in the air. Very few have ever pretended that they have witnessed the connection. A few years ago, I saw a statement naming two individuals who had witnessed it. As one of them was a perfect stranger to me, perhaps I ought to qualify my opinion, and say that it *is possible ;* but the other one

I happen to know is not perfectly reliable in all things, and if the truth in this case, is to be established upon *his* testimony, I fear it will lack support.

The queen upon her return, frequently bears evident marks of her connection with the drone, and usually begins to lay in two days afterwards, and continues throughout the season, unless some special interruption occurs.

NUMBER OF EGGS THAT A QUEEN WILL LAY.

The number of eggs that she will lay in twenty-four hours, is a subject on which all writers do not yet agree, probably owing to the fact that the number varies from one or two hundred to three thousand. Take a queen that has been reared in a small box, and she will soon fill all the combs after she begins, and when there is no occasion for laying many, there are less deposited. She remains small in size, and seems to adapt herself, partially at least, to the necessities of the colony; but this same queen, introduced to a strong colony with suitable combs, in a honey season, will, in less than a week, greatly increase in size, so as hardly to be recognized, and will deposit two or three thousand eggs daily. This statement, when first heard, is received with a very large margin by almost every one: "the thing seems impossible;" and yet a little patient observation convinces the most skeptical. I have had a colony in an observatory hive, where every egg deposited could be seen. Visitors have frequently counted the eggs as deposited, for ten or fifteen minutes, and all have estimated the number laid in twenty-four hours to be over three thousand. Mr. Harbison says: "During the past season I worked a number of queens to their full capacity for producing eggs, in strong colonies, by frequently changing combs from which brood had just emerged in artificial swarms where the queen had not yet become fertile, for combs stocked with eggs and larvæ, stimulating them constantly by keeping them well supplied with food, when honey abroad became scarce. I put two of these combs, being about twelve inches wide, by fifteen or sixteen deep, into a strong colony, where the queen was very prolific. Over two-thirds of the cells were empty when put in, and within four or five days they were all stocked with eggs, except a few that were stored with pollen.. This was by no means a single occurrence. It was repeated again, and again, making at least 10,000 eggs laid in four or five days."

A person desirous of approximating to the number of eggs deposited, without being able to actually count them, can make an estimate as follows. It will satisfy him that a queen is the mother: If we examine a thrifty stock in the height of the breeding season we shall find the combs filled with brood, amounting frequently to three-quarters of all in the hive. By observing the number of cells to the square inch, it is easy to get the number to the square foot, then multiply this by the number of the combs in a hive, and we shall have the whole number of cells. For example, a piece of worker comb one inch square contains about fifty cells, including those on both sides. At this rate, a piece twelve inches square contains over 7000. Suppose a hive contains eight such combs, and that 120 square inches of each comb are used for brood, we have eight times 120 square inches of brood—960—fifty to the square inch would multiply into 48,000 cells. One or two of these combs would contain cells for drones, which are a little larger, and the number would be thus somewhat reduced. Also some might be empty, the young bee having just left them, and a few here and there might be occupied with bee-bread or honey. Admitting the necessary deduction to be one quarter, we would have left 36,000 *cells actually occupied at one time with brood in various stages of development!* We must remember that the time from the egg to the mature bee, is not over twenty-five days at most; hence all that are in the cells now, must have been put there by the queen within the last twenty-five days! This gives an average of nearly 1500 per day!

It is common to find estimates that a single female will lay from 70,000 to 100,000 in a season. One says 90,000 in three months; most writers are apt to confound the number matured with the number laid. Let the number laid be what it may, thousands are never perfected. During the spring months, in medium and small stocks where the bees can protect but few combs with animal heat, I have often found cells containing a plurality of eggs, two, three, and occasionally four in one cell. These supernumeraries must be removed, and may frequently be found in the dust on the bottom board.*

Another portion of eggs is wasted whenever a supply of their food fails. If we remove the bees from a stock during a scarcity, when the hive is light, we will be very likely to find hundreds of eggs in the cells and only very few advancing from that stage toward maturity. I have thus found it in autumn, in July, sometimes the first of June, and in fact, at any time when the maturing of the brood would be likely to exhaust their stores, and endanger the supply of the family. Now instead of the fertility of the queen being greater in the spring and first of summer than at other times, as we are often told, I would suggest that a greater abundance of food at this season, and a greater number of empty cells, may be the reason that more bees are matured.

---

* This is a good test of the presence of a queen. Sweep off the board clean, and look a day or two after, for these eggs. Take care that ants or mice have no chance to get them; they are as fond of eggs for breakfast as any one, and might deceive you by removing the eggs. When any eggs or immature bees are found, no further proof of the presence of a queen is needed.

WHEN DRONES ARE REARED.

Whenever the hive is well supplied with honey and bees, eggs are deposited in the drone cells.

WHEN QUEENS ARE REARED IN SWARMING HIVES.

Also, at the proper season, when the hive becomes crowded with bees, and honey is plenty, the preparations for young queens commence. As the first step towards swarming, from one to twenty royal cells are begun, and when about half completed, the queen (if the conditions continue favorable,) will deposit eggs in them.* These are glued fast by one end like those for the workers. When hatched, the little worm is supplied with a superabundance of food; this appears from the fact that I have frequently found a quantity remaining in the cell after the queen had left. The consistence of this substance is about like cream, the color some lighter, or just tinged with yellow. If it were thin like water, or even honey, I cannot imagine how it could be made to stay in the upper end of an inverted cell of that size, in such quantities as are put in. Sometimes a cell of this kind will contain this food, and no worm to feed upon it. I surmise that the bees have compounded more than their present necessities require, and stored it there to have it ready; also, that being there, all might know for whom it was designed.

The taste is said to be "more pungent," than that of the food given to the worker, and the difference in food is assumed to change the bee from a worker to a queen. It can not be the shape of the cell, because I have known queens to be raised in cells that could not be distinguished from worker cells, by ordinary observers.

---

*I do not assert this positively. All my observations indicate it, yet I have never seen her in the act.

WHEN QUEENS AND DRONES ARE DESTROYED BEFORE SWARMING.

If from any cause, honey fails so far as to make the existence of a swarm in any way hazardous, the preparations are abandoned, and these young queens destroyed, in all stages from the egg to maturity. When an occurrence like this takes place, the drones next fall victims to the failure of honey. A brief existence only is theirs; such as are perfect, are destroyed without mercy, and those in the chrysalis state are often dragged out and sacrificed to the necessities of the colony. Such as are allowed to hatch, instead of being fed and protected as they would be if honey were abundant, are permitted while yet weak from the effects of hunger, to wander from the hive, and fall to the earth by hundreds.*

These results attend only a scarcity in the early part of the season. The massacre of July and September is quite different. The drones then have age and strength; an effort is apparently first made by the workers to drive them out without proceeding to extremes; they are harassed sometimes for several days, the workers feigning only to sting, but very few are dispatched in that way; yet there is evidence proving beyond doubt that the sting is sometimes used. Hundreds will often be collected together in a compact body at the bottom of the hive; this mutual protection affording a few hours' respite from their tormenters, who do not cease to worry them. In a few days they are gone, and it is difficult to tell what has become of them. If the hive is well supplied with honey in September, some of the drones have a longer lease of life given them. I have seen them as late as December. When the best hives are poorly supplied with stores, the bees will rear no drones the ensuing spring, until the

---

*The destruction of drones at this time does not prove that there will not be any swarms, as some have asserted; but it shows that there is a scarcity of honey, and that swarming is put off indefinitely, if not altogether.

flowers yield a good supply. I have known one or two years in which no drones appeared before the last of June; at other times, thousands are matured by the first of May.

THE OLD QUEEN LEAVES WITH THE FIRST SWARM.

The old queen leaves with the first swarm, and as soon as cells are ready in the new hive, she deposits her eggs in them, at first for workers, the number corresponding with the supply of honey and size of the swarm. When the supply of honey fails before the time for leaving the old stock, she will remain there, and continue laying throughout the season.

WHAT BECOMES OF THE BEES WHEN NO SWARM ISSUES.

As many bees die or are lost during their excursions, as are replaced by the young ones. In fact they often diminish in number rather than increase; so that by the next spring a hive that has cast no swarm is no better for a stock, than one from which a swarm has issued. We are apt to be deceived by bees clustering outside, towards the latter end of the season, and suppose it hardly possible for them to get in, when in fact it may be caused by hot weather, full stores, etc.

WHEN A YOUNG QUEEN TAKES THE PLACE OF THE OLD ONE.

In ordinary circumstances when a swarm has left a stock, the oldest of the young queens is ready to emerge from her cell in about seven or eight days; if no second swarm is sent out she will take her mother's place, and begin to lay eggs in about eight days. Two or three weeks is all the time in the whole summer in which eggs can not be found in all prosperous hives.

NUMBER OF DRONES.

The relative number of drones and workers, when the latter are most numerous, doubtless depends on the size of the hive. When a swarm is just hived, the first cells constructed are of the size for workers. If the hive be very small, and bees numerous, it may be filled before they are fully aware of it, and but few drone cells be built; consequently but few drones can be raised. If the hive be large, considerable honey will be stored, and cells for storing honey are usually of the size intended for drones, and these will be made as soon as the requisite number for workers is provided. It is said that more drone cells are made when the queen is quite old. An abundant yield of honey during the process of filling a large hive, would also cause a greater proportion of these cells to be built. The amount of drone brood being governed by the same cause, is also a strong argument against large hives, as affording room for too many of these cells, where an unnecessary number of drones might be reared, thus causing a useless expenditure of honey, etc.

THEORIES RELATIVE TO DRONES.

It is now determined that an egg deposited by an ordinary queen, in a drone cell, becomes a drone, and in a worker cell, a worker. I know that theories differing very materially from the foregoing, are advanced by nearly all writers. One says: "In spring the queen lays about two thousand eggs of males, resumes it again in August, but during the rest of the intervals she exclusively lays worker eggs. The queen must be at least eleven months old before she begins to lay the eggs of males." Dr. Bevan says, "the great laying of drone eggs usually commences about the end of April." All these theories are at fault. It is proved beyond dispute that drone eggs are laid at all seasons whenever the condition of the hive will warrant it. But there are those who have investigated farther, and who give us another theory: that the eggs for the two kinds of bees are produced separately, and that the queen knows when each kind is ready, and the workers know it also.

The fact that all eggs laid in drone cells will produce drones and nothing else, is to be accounted for. There is no possibility of setting this aside. The attempt to rear queens from such has so utterly failed with myself and others, that we have no longer any hope of success. The reason undoubtedly is, that eggs laid in drone cells are not impregnated. Queens with faulty wings, or otherwise unable to fly out to meet the drones, or such as are raised late in the season, when no drones exist, are certain to prove drone layers; every egg they deposit, whether in worker or drone cells, produces a drone. I have frequently, since obtaining the Italian, reared queens intentionally late in the season, that I may have drone-laying queens for the purpose of raising early drones. Such failed to meet the drones, and were drone layers in consequence.* Whenever the brood of the fertile workers has matured, it has proved to be drones.† No one will pretend that these

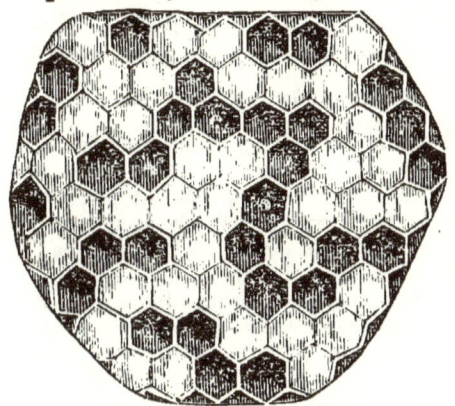

Fig. 5.—BROOD FROM A DRONE QUEEN IN WORKER CELLS.

---

*A drone queen, when laying in worker cells, does it more irregularly, or the bees do not nurse all that are laid. About half the cells are sealed over after being lengthened at least one-third. It has been recommended to " destroy such a queen and substitute another; and as the combs are worthless, destroy them, and let the bees build new." I have found these combs as good as new ones, and would advise retaining them.

†I never witnessed the phenomenon of a fertile worker until after I had been raising Italian queens in the small rearing boxes for some time. I had used clean drone comb in several of them, and in some that had been without a queen a long time. I discovered eggs in the cells. Some contained as many as six, put in rather uningeniously, as if it were the work of a novice. Some were sticking

have become impregnated. All this *indicates*, if it does not prove conclusively, that all drone eggs are unimpregnated. There is still another indication that they are not impregnated. The Italian queen that has met the native drone, and brings forth a mixed progeny of workers—half Italian and half native—will produce just as pure drones as her mother, or one that has never met the drone. Does not this militate against any theory that the vivifying influence is incorporated with the egg in its formation?

on the side, half way to the bottom, and others were on different parts of the bottom. Some of the cells contained larvæ pretty well advanced, and that eventually matured into apparently perfect drones. A day or two after, on taking out a comb, I found a worker in the very act of laying. Her abdomen was inserted its whole length, her head, thorax, and wings being all that was visible of her body She was not disturbed at all by the removal of the comb, but continued the important operation of depositing an egg, the gravity of her countenance indicating that she considered herself the important personage of the colony

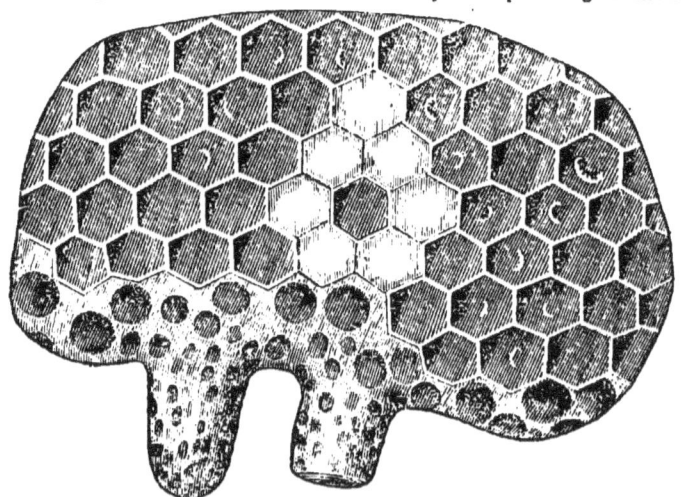

Fig. 6.—COMB SHOWING THE DIFFERENT SHAPE OF CELLS WHEN AN ATTEMPT IS MADE TO RAISE A QUEEN FROM DRONE BROOD.

by being elevated to the dignity of becoming the mother of a drone. I found that the length of time required by her to deposit an egg was three or four times greater than that usually occupied by a queen. A day or two after, I caught three dignified matrons *at one time* engaged in this all important and not-to-be deferred business, and afterwards observed several others thus occupied. I noticed that the phenomenon was usually produced by keeping the little colony

This principle is of immense value to all who would Italianize their apiaries.

To account for their not being impregnated, especially those laid by a perfect queen, Mr. Langstroth says:

MR. WAGNER'S THEORY.

"My friend, Mr. Samuel Wagner, of York, Pa., has advanced a highly ingenious theory, which accounts for all the facts, without admitting that the queen has any special knowledge or will on the subject. He supposes that when she deposits her eggs in the worker cells, her body is slightly compressed by their sides, thus causing the eggs as they pass the spermatheca to receive its vivifying influence. On the contrary, when she is laying in drone cells, as this compression can not take place, the mouth of the spermatheca is kept closed, and the eggs are necessarily unfecundated." Mr. Harbison replies that he has no faith in this "very plausible theory," and thinks that "facts, further experience and observation, will demonstrate its fallacy." It appears that it is easier for him to pull down than to build up, because, after showing the weakness of

some days longer without a queen, and that they must be provided with plenty of drone cells. I never knew them to lay in worker cells. The instinct that prompts the desire to preserve the colony from destruction, inspires efforts to which nature will not grant success. They even endeavor to rear queens from these eggs, on some occasions. The great wonder is, why a worker should lay at all. The only solution that I can offer at present is, that the knowledge of, or grief at the loss of their mother, changes the internal structure of a mature bee, and develops eggs sufficiently vitalized to hatch drones. The theory that worker layers were raised near a queen cell, and by accident were fed a little royal pap, will not explain it at all. These workers were taken from a colony that had never raised a queen, and they probably never thought of depositing an egg so long as the queen was present. If this great anxiety for the mother was any less, they might sometimes neglect to avail themselves of the means of providing one, when they had the power.

The phenomenon of other insects than the bee, producing young without direct impregnation may be witnessed in the aphis, (Plant louse.) Not only one, but several generations of females, are brought forth in succession. Towards the end of the season a few males are produced, which continue the species for a few months longer.

this theory, he offers one, that, to me, appears still more fallacious, and still more beset with difficulties. He states as objections to Mr. Wagner's theory, first: "that the abdomen of the queen, where the sac is situated, is. so small, that when thrust even to the bottom of the worker's cell, it cannot be sufficiently compressed to impregnate the egg as represented, in passing its mouth." Again: " When the old queen is hived with a swarm, she commences laying eggs as cells are ready, and often lays in worker cells when only one-eighth of an inch high. Is it possible that the abdomen of the queen receives any pressure from the sides of the cells whilst in the act of thrusting her ovipositor into the cell to deposit the egg?" He goes back to the theory of periodical drone-egg laying, not to the extent of limiting it to two periods in the season, but to certain periods. He says, in substance (I condense his remarks on this point), that he thinks it highly probable that the queen knows that an egg in a drone cell will bring forth a drone; knows when it is proper to raise drones, etc. Who ever saw eggs laid in drone cells in mid-winter or early in spring, until nearly time for swarming? Yet all strong stocks raise brood from January until summer. He has " cut holes in a worker comb and inserted corresponding pieces of drone comb which remained empty while all around would be filled with worker brood, etc., proving that no drone eggs are laid until the general simultaneous laying of all strong colonies. He gives us his concluding theory which I quote at length.

MR. HARBISON'S THEORY.

"At present, I shall content myself with believing that a sufficient portion of the seminal fluid to cause the egg to germinate, is incorporated with it in its formation. The eggs to produce drones or males are generated in, or produced from one side or branch of the ovaries, and those producing females from the other side. We find that the ovaries are separated into two equal parts, (according to Swammerdam, after whom Langstroth copies,) having no connection whatever, except that the contents of each branch is discharged through the common oviduct or passage. Over the outlets of the passages or oviducts opening from each of these divisions into the main channel or common oviduct, the queen has full control, and fully understands that eggs from the one division will produce drones, and from the other, workers; and the anomaly of drone-laying queens arises from the imperfect development of that part of the ovaries which produces eggs for workers."

Let us see what there is in support of the two theories. He believes that there is enough seminal fluid to cause the egg to generate, incorporated with it in its formation. Yet in another place he admits that the queen is provided with a receptacle for the male element, and doubts that any eggs are ever hatched that are not impregnated direct, in the following language. "That this is true, permit me at present to doubt; its assumptions are too extravagant, and so far from harmonizing with all animated nature, with which I am in any way conversant, etc."

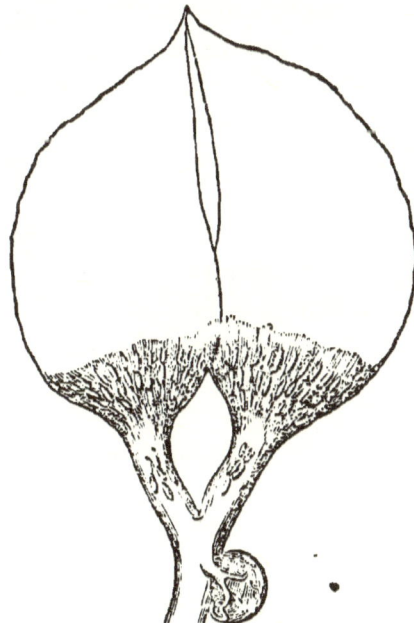

Fig. 7.—OVARIES OF THE QUEEN.

He should remember that we start with some facts not in harmony with any thing "with which we are conversant." Where in the whole round of animated nature is there a female producing hundreds of thousands of eggs in a life time? We have heard of a kind of ant that exceeds even this, but we are not familiar with it. In animated nature, as far as our observation extends, eggs are usually produced in series of dozens or hundreds, and the male is met for each brood. A female may produce one brood or more, but the queen will, under some circumstances, continue an uninterrupted series from January to October. One impregnation is operative for a life time. Is there not wisdom in thus economizing the vital fluid, and using it only when necessary?

But eggs of some animals are impregnated after their formation, even after they are discharged, as is the case with those of most kinds of fish, frogs, etc. The providing of the sac or receptacle of the queen, is an admirable display of the wisdom of the Creator, in adapting means to ends. Mr. Harbison's supposition that because the ovaries of the queen are separated into two equal parts, that when the queen produces only drones, one side is imperfectly developed, does not enlighten us much, even if we admit it. I see no reason why the side in which the drone eggs are formed, should not sometimes be imperfectly developed as well as the other, and we occasionally have a queen that lays no drone eggs; such a case has never been reported, to my knowledge. That she "fully understands that eggs from one division will produce drones, and the other, workers," I can not comprehend any better than that she would understand equally well when a drone egg was about to be laid, if they were all formed in one mass. If drone and worker eggs were separated in two divisions, it would seem that the mass of each would be proportionate in size to the quantity laid of each; but they are represented as alike in size. Perhaps no one would dispute it if I should say that hundreds of one kind are laid to one of the other, on an average. Mr. H. says that a queen will lay none but worker eggs from January until about the swarming time, when simultaneously all queens in strong colonies lay drone eggs. Is one division idle all this time? A queen in a weak colony, with but little honey, may be kept all summer without raising a drone; they are frequently so kept. What becomes of the division of drone eggs all this time? If it were natural for her to deposit them at a particular season, like the putting forth of buds and leaves, then the queens of small families should produce drones according to the season, and not according to the condition of the family.

That the queen knows any thing about the kind of eggs she is depositing, is, to me, very doubtful. One circumstance witnessed by myself, has a strong bearing on this subject. A colony in a glass hive had remained quite small up to the middle of June. One outside sheet of comb was three-fourths worker cells. For several days I had observed the bees cutting off the cells in which they had stored honey the previous season, for the purpose of rearing brood. I soon discovered eggs in both worker and drone cells. Shortly after, on opening the door, I found the queen engaged in laying eggs in the drone cells. Nearly every one already contained an egg. Most of these she examined, but did not use them. Six or eight, only, appeared to be unoccupied, and in each of them she immediately deposited an egg. After depositing the last one in the drone cells, she continued to search for more empty cells, and in doing so, she passed, apparently by accident, upon the worker cells, where she found a dozen or more empty, in each of which, without hesitation, she laid an egg. The whole time occupied was not more than thirty minutes. Each produced according to the cell used, drones in drone cells, and workers in worker cells. It is hardly worth while to tell me that she knew her series of drone eggs was exhausted just when she laid the last one in the drone cell, because I should at once inquire why she examined so many of those cells, if she did not intend to use them, had they been empty. It will require very positive evidence to convince me that those worker cells did not receive the identical eggs that would have been placed in drone cells had they not been pre-occupied.

But can we account for the impregnation of some, and the non-impregnation of other eggs laid by the same queen in the space of half an hour, aside from the theory of Mr. Wagner? Among the objections raised to this by Mr. H., the strongest appears to be that the eggs laid in cells one-eighth of an inch deep, are just as sure to produce workers as those laid in deeper ones. It is considered by some persons as entirely fatal to the whole theory. For myself, I feel very reluctant to admit that it affects it in the least, yet I must acknowledge that it is somewhat obscured by the circumstances.

I very much hope that something explanatory will yet be discovered, because if it is rejected, there are so many things favoring this theory that will have to be otherwise explained, that the task will be very heavy. It is possible that when we have an arrangement by which we can witness the depositing of those eggs in such shallow cells, we shall discover something that will shed more light upon the subject. It may be, that, just at the moment of the passage of the egg, or the act of laying, the contents of the abdomen are crowded downward, and it enlarges sufficiently to touch the sides of a cell only one-eighth of an inch deep.

When I first saw the smallest queen that I ever raised, whose body was even smaller than a worker's, it occurred to me at once that if she ever laid, it would be a test of this principle. Her body being small it could not of course be compresssed like others, and a large portion of her progeny would prove to be drones in worker cells. The result was just what was expected—one half were drones. (This queen was lost on introducing her into a full colony.) I have noticed, and no doubt others who have raised queens in the small boxes also have, that from the first eggs that the queen deposits, before her body is enlarged by the accumulation of eggs, there are many drones, even in the worker cells.

There are two sides to this question. The queen either knows when she is about to lay drone eggs, or she does not. If it is admitted that she *does*, another question immediately arises, do the workers know it also? Whenever the condition of the colony is such that drones may be wanted, we find them preparing for them. If they have no cells made, and there is room in the hive, they construct them; if they are made, they cut them down, if they had been used for honey, and otherwise prepare them for the eggs. Do they do this because the queen has imparted to them the knowledge of her wants just then, or is it the result of common instinct? The hive, at such times, is sufficiently populous for the bees to cover the comb and maintain the requisite heat. They are getting a supply of honey from the flowers, and simultaneously all good stocks rear drones. The stimulus of obtaining the honey seems adequate to produce the result. It is not necessary that the honey should be obtained from the flowers at the time. Sometimes it may have been stored the previous year, or a large quantity may have been fed, and then strong colonies will rear drones a month in advance of the season. We can stimulate a strong colony to rear drones throughout the season, even as late as October, by keeping up sufficient warmth, and a liberal supply of food. I have frequently raised Italian drones out of the honey yielding season, when the natives were mostly destroyed, for the purpose of serving queens reared out of season. There is then much less risk of their meeting the black drone. There is still another theory of this matter of sex, offered by Mr. E. Kirby, but as I am unable to comprehend it fully, I will not undertake to explain it to others.

Twelve years ago I dismissed this knotty subject with this remark, "I shall leave this matter for the present, hoping that something *conclusive* may occur in my experiments, or those of others. At present I am inclined to think that the eggs are all alike, but am not fully satisfied." Since then we have advanced somewhat, in theories at least. We have facts pointing very clearly to the conclusion that eggs producing drones are not impregnated. By patient perseverance, I trust that there will yet be more light thrown upon this interesting subject.

I am aware that this matter is of but little interest to many readers, and I am advised to adhere to the plain and *practical*, and avoid speculative topics. I shall endeavor to do so generally, but this is, to me, of such special interest, that I could not well avoid devoting a little more attention to it than will be agreeable to all.

CHAPTER III.

HIVES.

SATISFACTION IN HAVING NO PATENT.

There is a satisfaction, in being able to express my views on a subject involving so many conflicting interests, and feel that no one can accuse me of selfish motives. I have kept clear of all interest in the patent swindles of the day, and have refused tempting bribes for a simple endorsement of some particular "pattern bee-hive." I have refrained on principle from inflicting another patent on the community, whereby I doubtless could have made some hundreds of dollars; for all the different methods of constructing a bee hive are by no means exhausted, neither is the race of credulous bee-keepers extinct. I have put myself in antagonism with the patent-vender, have endured his abuse, his sneering ridicule, and unfounded accusations. For what? Certainly not in the hope of any pecuniary reward. It is said that he who causes two blades of grass to grow where but one grew before, is a public benefactor. So it may be said of the man, who even indirectly, aids in saving a portion of the inestimable sweetness now wasted on the air, for want of proper means to secure it. I have ever been anxious to advance apiarian science, and promote the interest of the apiarian, and if I succeed in ever so small a degree I shall feel amply rewarded.

How can a man judge of the requisites of a bee-hive, unless he is thoroughly versed in the natural history of bees? Not one in a hundred of those spreading patent hives broadcast over the land, is capable of giving an intelligent opinion concerning the habits and requirements of bees. A patent is based upon some peculiarity of construction, by which some real or fancied convenience is obtained, and thenceforth that convenience is proclaimed to be the one thing needful for a bee-hive, although it may supplant other and more desirable qualities, if it does not induce some positive evils.

NO DIFFICULTY IN OBTAINING CERTIFICATES AND PREMIUMS.

There is no difficulty in getting certificates of the enormous quantities of honey produced by each particular hive. With few exceptions, all patent-venders are provided with them, as well as with "premiums" for "best bee-hive," received at all the Town, County, and State Fairs, throughout the country. A premium from our Fair Committees is no longer a recommendation,—every thing is recommended as well as patented; and when a man comes along who has nothing to offer in favor of his hive, further than the unprecedented amount of honey secured by it, and a favorable notice from a stupid committee, I feel very much like dismissing him without ceremony; it is evident he has chosen a sorry vocation.

HIVES FURNISHED FOR TRIAL.

I have, during my bee-keeping experience, received a score or two of patent hives, with the right to use, and a request that I would give them a trial. Some patentees were no doubt sincere in the belief that I would find their hive the "ne plus ultra" of all contrivances, while others, less honest, were evidently only seeking a word of commendation, which would go far towards establishing their humbug in the confidence of the easily deluded public. It is often unpleasant to refuse so simple a favor as a trial of a hive, but although furnished gratis, there are seldom good points enough about them to make it worth the trouble; and further, it is not desirable to have many different patterns in one yard. None but the experienced can realize the importance of this last consideration, especially where movable combs are used. Besides, many of these hives are a positive damage to the bee-keeper. I think it will be an easy matter to show that when one desirable point has been gained by a departure from simplicity, it is usually attended by a corresponding evil.

NECESSITIES OF THE BEES.

All variations from the simple box are for the benefit of man, not of the bees. The wants of the bee are few and simple. A suitable cavity for the combs is all that is required. In good seasons, instinct will prompt the collection of a greater supply than is needed for winter. I will guarantee that more honey will be stored in a barrel, box, or hollow log, just large enough to hold all that is gathered, than in any patent fixture ever presented for this special purpose. It is just as rational to contend, that, with the same facilities, bees will store more in *your* barrel than in *mine*, as that they will store more in one patent hive than another.

GREAT DISCOVERY FOR PATENT-VENDERS.

When honey is stored in the apartment where brood is raised, it is liable to be mixed with occasional cells of pollen, and cocoons left by the young bees. The discovery that by making a division in the hive, that part separate from the brood would be free from all impurities, opened an ample field for speculation, and different methods of making the necessary division were at once invented. The chamber hive was probably the first of the kind.

PRINCIPLES OF DIFFERENT HIVES.

Then, to prevent the depredations of mice, the suspended hive was contrived. Soon after, the inclined bottom board was added to throw out the worms. When it was discovered that bees destitute of a queen would rear one from eggs destined for workers, dividing hives of various forms were at once presented. Comb used a great many years becomes thickened and black, and needs changing; hence the changeable hive. "Non-swarmers" have been introduced to save risk and trouble. "Moth-proofs" are offered to prevent depredations of the moth, etc., etc. I will examine some of the principles upon which these are founded, and then give my views of a good hive.

CHAMBER HIVE.

The chamber hive is made with two apartments, the lower and larger for the permanent residence of the bees,

3

the upper or chamber for the boxes. Its merit is, that the chamber affords all necessary protection for glass boxes, and is a permanent cover. Its demerits are: it is inconvenient to handle, it occupies too much room when put in the house for winter, and only one end of the boxes can be seen when on the hive. They can not be properly examined without taking them off, and thus disturbing the bees.

SUSPENDED HIVE.

The suspended hive may effectually exclude the mice, and answer all the purposes for which it was designed, but there are evils originating in this very advantage that may counterbalance it. The inconvenience of inspecting the hive at any time may induce a habit of neglect that is often fatal to success. When all the cares of an apiary are as light as possible, there is too great neglect of duty; hence necessary attention should be as little burdensome as possible.

INCLINED BOTTOM BOARD.

The inclined bottom board for rolling out the worms—the basis of several patents—may be said to be an utter failure. Worms are not disposed of so easily, for when one drops from the comb, if it ever does, it has a thread attached above, by which it may climb to its former position. Should it be dead when it falls, or so cold that it can not spin a thread, a strong wind might shake it off—and what then ? The objections to this are the same as to the suspended hive. They are a damage to whoever uses them, aside from the expense of right and construction.

DIVIDING HIVE.

The dividing hive was constructed to multiply stocks at pleasure. The fact that bees would rear a queen to replace one lost, gave rise to some very wild speculations. To make a hive in two parts, and when full, separate them, making two, and then put an empty half with each full one, on the presumption that the portion without a queen would rear one, was a theory that seemed very well until put in practice. I made a hive of this kind, and a Mr. Jones, a little later, did the same, and obtained a patent, but when they came to be put to the test of practice, we were taught a lesson. A medium-sized swarm put into such a hive will first fill one side down with nearly all brood combs, and this apartment will, most of the time, afford all the room needed for breeding. When they commence in the other, they will build store-combs, the cells being too large for rearing workers. A swarm large enough to fill both sides at once will do better, but it will construct more store-combs than are profitable. In many cases when the colony is divided, the result will be no brood in one apartment from which to raise a queen, and a strong probability that the old queen is with the brood, and the part without her must therefore run down. If by chance there is sufficient brood from which to raise a queen, so small a part of the comb is fit for breeding that they can raise but few bees, and the colony will remain weak and thriftless for a short time, and then die. I also found that a colony would often starve with abundant stores. Bees take up their winter quarters among the brood combs, in the apartment where there is but little honey; if it is all exhausted during protracted cold weather, they must starve. Only frequent intervals of warm weather, or warm winter quarters, can avert such a fate.

CHANGEABLE HIVE.

The very kind effort to prevent the bees from becoming dwarfs, has given rise to many forms of the changeable hive. We all know that when the young bee first hatches from the egg it is nothing but a worm, that it is fed a few days, and the cell containing it sealed over with a waxen covering. It then spins a cocoon, or lines its cell with a coating of silk, inconceivably thinner than the thinnest paper, which remains after the bee leaves the cell. It is evident, therefore, that after a few hundreds have been reared in a cell, each one leaving its cocoon, such cell must be somewhat diminished in size, and after a time become so small that the young bee cannot attain to its proper size. It therefore needs to be removed that the bees may replace it with one of full size. This is all very consistent, and were it not that the patent-vender takes advantage of the bee-keeper, through ignorant or designing misrepresentation, I might have but little to say on this point.

The most simple form of this class of hives consists of several stories one above another, with holes or cross-bars for communication, each section or story forming a hive five or six inches deep. Every year a full one is taken from the top, and an empty one added at the bottom. As there are usually about three, they are changed every three years, consequently none of the combs are over three years old. The one taken off usually contains the most honey. But of what quality is it? It is even inferior to some pieces that may be selected from the box hive. Every comb and nearly every cell must, at some time, have been used for breeding, consequently they contain either cocoons or pollen, and are not desirable or fit for the table until strained. A hive of this class is one of the worst in which to winter bees. It is objectionable on the same score as the dividing hive—bees in one part and honey in another. Every hive on this principle is open to the same objection, whether the sections are placed one above another, or upright, side by side. Hives of this kind may be considered the most pernicious of any. They rob us of a goodly quantity of surplus honey, compel us to hazard greater risk in wintering, and cause the consumption of several pounds of honey for the renewal of a portion of the comb every year. And what are our returns? In their most prosperous condition, some thirty pounds of inferior hive honey. When *properly managed,* this same colony would probably store a much larger amount of *pure* surplus, the market value of which would be ten times greater than that of the other. It will not do to reckon the value of *new combs* as an equivalent for all this sacrifice. I can assure the reader that there is no profit in such frequent renewal of the combs. All experienced and disinterested bee-keepers will bear testimony to this. Bees hatched from combs used for breeding a dozen years, are not dwarfed enough for the difference to be perceived. The bees seem to make a provision for this emergency by making the sheets of comb a little farther apart than necessary at first, and the diameter of the cell a little greater than the young bee requires. The angles of the cells fill up in time, and as the bottom fills up faster than the sides, the bees add a little to the length, until the ends of the cells upon two parallel combs approximate so closely that the bees can not pass freely; before this time it is unnecessary to remove combs on account of age.

I find it estimated by writers that twenty-five pounds of honey are consumed in elaborating about one pound of wax. This may be an over-estimate, but no one will deny that some is used. I am satisfied from actual experience, that every time the bees are obliged to renew their brood-combs, they would make from ten to twenty-five pounds of honey in boxes; hence I infer that their time may be much more profitably employed than in constructing brood combs every year.

Now, to have the bee-keeper deluded into the belief that by paying for the privilege of injuring his bees, he is benefiting himself, is too much for ordinary patience. I have said nothing about the expense of construction, which is, at least, three times that of common box hives, and it is nothing but a box hive after all. This item alone is worthy our attention.

VENTILATING HIVE.

In cold weather, bees throw off moisture that lodges on the combs and sides of the hive, and causes mold. The patent-vender is at hand with several specifics for getting rid of it. The most effectual that I have seen—Mr. Furlong's—is a hive with cross-bars at the top to support the combs, and panes of glass set up like the roof of a house, on which the moisture condenses, and runs down into a little trough of tin, which conducts it outside of the hive. This hive is much more tolerable than the dividing hive just mentioned, as this method of disposing of the moisture is preferable to the open holes. Were it not for the fact that the same result can be secured quite as effectually at far less expense, this hive might be desirable. (See description of box hive with straw mat for top in chap. XXIII.)

MOTH-PROOF HIVES.

To keep the worms from the hive, has exercised the ingenuity of our accommodating gentlemen of the patent fraternity, for a long time, and they " have succeeded beyond all expectations." The noticeable feature in men of this stamp is, that the less they know about bees the more they presume to teach others. In fact, one who is at home on the subject, does not believe a word of their professions. He sees well enough that a moth can go wherever a bee can go, and that when the bees are gone, or too weak to drive them away, the worms are present. The worms can not destroy a *strong* colony of bees, especially if Italian, although the hive may be the most rickety old box imaginable, with hiding places for worms on every square inch. Put this by the side of the best finished "Moth-proof," and the chances are that the moths, or rather worms, will dispose of the latter first.

NON-SWARMERS.

A perfect non-swarmer has not yet been constructed, although we often hear it talked about. I heartily wish that one could be devised which would answer the requirements, and furnish the surplus in good shape for market. I have offered $100 for one that would not fail in more than one instance in ten. It is not forthcoming, however, showing that those who talk most of their ability to invent, have no confidence in their own profession. The only place in which one can put bees and not expect them to swarm, is a small dark room, and a few have been known to swarm even then. But here, the surplus is made on the outside of the hive, and is of unequal thickness, and in all shapes, thus being unsuitable for market. I have tried the experiment of putting on boxes, as on other hives, but they seem to ignore them entirely, making combs at random on all parts of the hive.

When a person wishes to keep a few bees for the sake of the honey for home use, and wants the *least possible trouble* with them, he will probably be satisfied with this hive. But if he expects to sell a few thousand pounds, he does not want it in such an unsalable shape. I contrasted the profit of such a hive, with that of a swarming hive, in the first edition of this work, but I made one mistake, of which an interested party has taken the advantage, giving an unfair representation to show the non-swarmer the most profitable. Instead of comparing a swarming hive with a true non-swarmer, placed as I have represented, he assumes a hive to be such, when it occasionally fails to swarm, and estimating the surplus in boxes that a good hive would yield, he contrasts this profit with what I gave as about one-third of the average yield of a swarming hive in good seasons. He simply changed the question to one not under consideration at all. I mention this to show that the statements of interested parties should be received with caution. I shall now make a similar estimate, but to prevent similar misrepresentations, I will give a little nearer the true yield of a swarming hive. Recollect, I speak of the real non-swarmer, in a small dark room. We start with one hive worth $5; at the end of ten years it is worth no more. The chances of its failing before that time we will not take into the account. We get annually, say, $5 worth of surplus—it will not be likely to be worth more, considering the shape it is in. This, with the value of the hive, will amount to $55. We will suppose that the swarming hive throws off one swarm annually, and stores $5 worth of surplus. To be moderate, we will call the average $2. The swarms will sometimes store $20 worth, but we will call their surplus worth $2 each. Commencing with an old hive that gives an increase the first year, at the end of ten years we have 1024 hives. These at $5 each, are worth $5,120. At $2 each, the surplus brings $4,092, which, added to the value of the hives, gives a total of $9,212, as a result to compare with $55. To prevent any misapprehension on this point, I will state that in this illustration, I do not intend to be understood that any one will realize such a profit, but it serves to show the relative advantages of swarming and non-swarming hives.

It is said that many of the pretended "non-swarmers" can be converted into swarmers, in two days, at the option of the apiarian. Colton could place on his hive six large boxes at one time, containing nearly 3,000 cubic inches. By removing these at any time when there were bees enough to fill them all, the room was so much contracted that the swarm was forced out at once. This was found to be more theory than fact, when put in practice. Bees do not generally swarm without previous preparation of at least a week. This hive, in *one* respect, is better than most patents. The large amount of room in the boxes, which might sometimes tend to prevent swarming, gives all the bees an opportunity to labor when they do not swarm, and consequently more surplus honey is stored on such occasions. I understand that the "Farmer's Hive," patented by Mr. Hazen, is on the same principle. Any amount of room in a hive *will not prevent* swarming. If they fill from 1600 to 2000 cubic inches with combs the first season, they will swarm the next, nine times in ten, if the season is favorable, without adding any new combs, although there may be ample room for them. To test this, I placed under five full hives of 2000 cubic inches, as many empty ones of the same size, without the top. I had a swarm from each. Only two had added any new comb, and these but little, showing that ample room will not prevent swarming. These hives swarmed when there was room to make comb—some before any was commenced, others just afterwards. Therefore, it is idle for any one to flatter himself with a prospect of success in such experiments, without an entire change in the conditions.

When a very large hive has been provided for a double, or extra large swarm, and they fill it the first season, they seldom swarm. They seem to have sufficient room in the large number of combs ready made, for all they can do, and there is no necessity for their emigration. The tyro asks what becomes of the bees raised in the course of several years. The answer in full will be found in another chapter, and I will only notice here, that after a certain maximum number is attained, there is no farther increase. They gain nothing in number from one May until the next.

I believe that with the exception of the Movable Comb Hive, I have now noticed all the principles worthy of attention, involved in patent hives. I will now speak of a class of hives that will pay better when put in use.

COMMON BOX HIVE.

I will first notice a hive in the simplest form. It has been called the "Quinby Hive," because it was the only one recommended in the first edition of this work, but the title has not always been given in a complimentary spirit. I have no claims whatever to this hive, as it was made and used long before my day. I recommend its use with some little alteration, but it is no more a "Quinby" hive than two or three others that I intend now to recommend. I have studied for years to secure the greatest amount of profit with the least possible expense, so that when I obtained five or ten dollars worth of honey, I need not pay it all for the hive and its appendages. I would keep a few colonies for amusement and instruction alone, but when I increase the number to hundreds, it is with the sole intention of making money out of them. I presume that very many of my readers are actuated by similar motives. I shall again recommend the box hive as the best and most economical for a large proportion of bee-keepers—those who have no interest, time, or patience to study the science of bee-keeping—till they can give a philosophical reason why they should use a different hive. One desirable feature about this hive, is, that no one has to pay for the right of using it.

PROPER SIZE OF HIVE.

After deciding upon the kind of hive, the next important point is the size. Dr. Bevan, an English author, recommends "eleven and three-eighths inches square, by nine deep, in the clear," making only about 1200 inches, and requiring so little honey for wintering bees, that when I read it, I found myself wondering if the English inch and pound were the same as ours. Whatever his experience, I think this size too small for bees in any place. We must remember that the queen needs room for all her eggs, and the bees need space to store their winter provisions; for reasons before given, these should be in one apartment. When this is too small, their supply of food is liable to be exhausted. The swarms from such hives will be smaller, and the stock much more liable to accidents. Yet I can imagine how one can be deceived by such a small hive, and recommend it strongly, especially if patented. Suppose you locate a large swarm in a hive near the size of Dr. Bevan's; the bees will occupy nearly all the room with brood combs. If you put on boxes, and as often as full, replace them with empty ones, the amount of surplus honey will be great; a very satisfactory result for the first summer, but in a year or two your little hive is gone. As we enlarge our hives, this result is modified, until we reach the opposite extreme, which is equally undesirable. If too large, more honey will be stored than is required for their winter use, of which it is evident that a portion might have been secured, had it been stored in boxes. Swarms issuing from such hives will not be proportionably large, and issue but seldom. They are of but little profit, in surplus honey or swarms, but have the advantage of being long lived.

Between the two extremes, as in most other cases, is found the correct medium. A hive 12 inches square inside, containing 1728 cubic inches, has been recommended as of the best size. This, I think is large enough in many sections, as the queen probably has all the room necessary for depositing her eggs, and the swarms are more numerous, and nearly as large as from much larger hives; there also is room for honey sufficient to carry the bees through the winter, at least in many sections south of 41° where the winter is somewhat short. This size will also do in this latitude, (42°) in some seasons, but not at all in others. Not one swarm in fifty will consume 25 pounds of honey through the winter, that is, from the last of September to the first of April. The average consumption in that time is about 18 pounds, but the critical time is later, about the last of May, or first of June, in many places. In latitude 42° and 43° they commence collecting pollen and rearing their young about the first of April; by the middle of May all good stocks will occupy nearly if not quite all their brood-combs for this purpose. But little honey is obtained before fruit blossoms appear, and when these are gone, no more of any amount is collected until the appearance of white clover, some ten days later. If during this season of flowers of fruit trees there should be high winds, or cold rainy weather, but little honey is gathered, and our bees have a numerous brood on hand that *must be fed*. In this emergency, if no honey remains from the stores of the previous year, a famine ensues; they destroy their drones, perhaps some of their brood, and for aught I know put the old bees on short allowance. This I do know, that sometimes whole families have actually starved at this season. This, of course, depends on the season; when favorable, nothing of the kind occurs. Prudence, therefore, dictates a provision for this emergency, by making the hive a little larger for northern latitude, permitting the storage of more honey, to take them through this critical period. From a series of experiments I am satisfied that 2000 cubic inches inside is the best size for this section. On an average, swarms from hives of this size are as large as any. The dimensions should be uniform in all cases, whatever size is decided upon. It is folly to accommodate each swarm with a hive corresponding in size; a very small family this year may be very large next, and the contrary the year following. A queen belonging to a small swarm is capable of depositing as many eggs as one belonging to a very large colony. A small colony which is able to get through the winter and spring, may be expected to be as large as any, another season.

DIRECTIONS FOR MAKING BOX HIVES.

Select one-inch boards of the proper width to make the hive about square, of the desired size, say 12 inches square inside, by 14½ deep. I prefer this shape for the box hive, but it is not all-important. I have had some 10 inches square, by 20 in length; they were awkward looking, but I could not discover any difference in the prosperity of their occupants. I have also had them 12 inches deep by 13 square, with the same result. A neighbor has used them 12 × 18, and 10 inches deep, with much satisfaction. One-third more room could be obtained for boxes, with this shape. In seasons when no swarm issues, the great number of bees present would thus find employment. If we avoid extremes, and give the required room, the form can make but little difference. It has been advised to plane the boards for hives, " inside and out," but bees when first put in such a hive, experience much difficulty in holding fast until they get their combs started, hence this trouble is worse than useless. When hives are not painted, the grain of the wood should never run crosswise, having the width of the boards form the height; not that the bees would have a dislike to this, but nails will not hold firmly, and will draw out in a few years. The size, form, materials, and manner of putting together, are now, I think, sufficiently understood. Sticks half an inch in diameter should cross each way through the centre, to help support the combs. A hole about an inch in diameter in the front side, half way to the top, is a great convenience to the bees coming home heavy laden. It is also essential when the hive is set close to the board, on account of robbing. It is likewise necessary to lower the hives to confine the animal heat as much as possible, when the bees are engaged in rearing young brood in cool weather, as warmth is necessary to hatch the eggs and develop the larvæ. Those who desire it, can make an additional entrance to the hive, by boring a few holes in the side close to the bottom.

TOP OF HIVE NOT FASTENED.

Instead of nailing a top to the hive, as I have heretofore recommended, with holes through which the bees may ascend to the boxes, I would suggest that there be slats across the top to support the combs, about three-fourths of an inch wide, by half an inch thick, and half an inch apart, one quarter inch below the top of the hive. Four or five strips, one quarter inch square, laid at equal distances crosswise the slats, will be just even with the top of the hive. The surplus boxes can be set on these, and the bees will find their way into them sooner than through holes in a top board. The queen is more liable to go up and deposit eggs, but not quite as much so, as if the boxes were directly on the slats, and there is not much risk after the hive is about full of comb, before the boxes are added, which it should be. If such hive is to stand in the open air for the winter, it will admit of a straw mat on the top, after the boxes are off, or the cap may be packed full of hay, straw, or corn-cobs, to receive the moisture.

A box for a cover or cap, 14 inches inside, will fit any hive. The height of this cap should be 7 inches. Of course other sizes will answer, but if we commence with one that we can adhere to uniformly, no vexations will arise by covers not fitting exactly. Where a double tier of boxes is used, covers must be made to fit. This cover, when on the hive, may rest on a strip of wood three-fourths of an inch square, nailed around on the outside, one inch below the top of the hive.

BEST SURPLUS HONEY BOXES.

Having told how to make the hive, I will give some reasons for preferring a particular kind of boxes. I have taken great quantities of honey to market, put up in every style, such as tumblers, glass jars, glass boxes, wooden boxes with glass ends, and boxes all wood, and have found the square glass boxes to be the most profitable. The honey in these appears to very good advantage, so much so that the majority of purchasers prefer to pay for the box at the same rate as the honey, to taking the wood and having the tare allowed. This rate of selling boxes always pays the cost, while we get nothing for the wooden ones. Another advantage in this kind of boxes is that the progress can be watched, and the boxes removed as soon as filled, thus preserving the purity of the combs.

DIRECTIONS FOR MAKING THE HONEY BOXES.

Select thin boards of pine or other soft light wood, dress down to one-fourth of an inch thick, cut the pieces for the top and bottom of the box, twelve and three quarter inches long, and six and three-eighths wide. Bore a row of holes in the center of the bottom. If the top of the hive is a board with holes through, make those in the box to match. Next get out the corner posts, five-eighths of an inch square, and five inches in length. For receiving the glass, cut with a thick saw a channel lengthwise on two sides, one-fourth of an inch deep, and one-eighth inch from the corner. A small lath nail through each corner of the bottom into the posts will hold them. It is now ready for the glass. Get 10×12, cut them through the centre, the longest way for the sides, and again the other way, five and five-eighths long, for the ends. These can now be slipped into the channels of the posts, the top nailed on like the bottom, and the box is complete. Boxes one-half or one-third this size are preferred by many customers, but the bees will store more honey in large than small ones. I have a method of holding the glass in place by means of pieces of tin, but it has so little advantage over those just noticed, that it is hardly worth while to describe it minutely.

GUIDE COMB.

It will be found of great advantage, previous to nailing on the top, to stick fast to it guide-comb, in the direction you wish the bees to work. This will also induce them to commence work several days sooner than if they had to start the combs themselves. Put in as many as you wish combs in the boxes. Pieces an inch square will do, and two inches is about the right distance apart. To fasten them, melt one edge by the fire, or melt some bees wax and dip one edge in that, and apply before it cools. For a supply of such combs, save all empty, clean, white pieces when removing combs from a hive.

For home consumption the wooden box answers equally well for obtaining the honey, but gives no chance to watch the progress of the bees, unless a glass is inserted for the purpose, which will need a door to keep it dark, or a cover over the whole like the one for glass boxes. Wooden boxes are generally made with open bottom, and set on the top of the hive. A passage for the bees directly from the box to the open air is unnecessary, and worse than useless. They like to store their honey as far from the entrance as possible. Unless crowded for room, they will not store much in the boxes when such entrances are made. Whether we intend to consume or sell our surplus honey, it is as well to have the hives and covers made so that we can use glass boxes when we choose.

When jars, tumblers, or other glass vessels are used, it is *absolutely necessary* to provide as many guides as you wish combs made, or secure a piece of wood inside, as they seldom commence building on glass, without some such inducement. The reader may have seen paraded at our fairs, or in the public places in some of our cities, hives containing tumblers, some of them neatly filled, others empty, with the magic sentence written upon them "*Not to be filled*," as if they were pretending to govern the bees by mysterious incantations, as a juggler sometimes performs his tricks.

I have termed the cap or box, a cover, but this should also be covered, with a board, if nothing else. A good roof for each hive can be made by fastening two boards together like the roof of a building; let it be about 18x24 inches; being loose, its position can be varied in accordance with the season. In spring, let the sun strike the hive, but in hot weather let the roof project over the south side, etc. The boxes described, can be used on any of the hives yet to be mentioned.

Fig. 8.—ROOF.

SOME DESIRABLE THINGS NOT FOUND IN BOX HIVES.

Every bee-keeper has found that there are several things desirable in a bee-hive that the makers of many improved hives never think of. He has seen stocks most promising in spring, containing the brightest combs, just the right amount of stores, and a strong colony of bees, begin to dwindle without any apparent cause, and has wished for some means by which he could inspect the interior, and ascertain whether the queen was lost or barren, or the brood diseased. He has often, in autumn, had colonies with too little honey for winter, and at the same time other hives with an over supply, and would like to be able to transfer some of this surplus to the light stock.

He has wished for an increase by swarming, and his bees have remained clustered outside the hive, refusing to swarm, the whole summer. He would welcome any invention by which he could divide them safely and profitably.

His bees would over-swarm, sending out many small ones not worth hiving, and ruining the old stock. How could he remedy this evil? He has found some swarms constructing entirely too much drone-comb, making the hive unprofitable ever afterwards, from the multitude of drones reared. How desirable sometimes to substitute worker for drone comb, and make it a profitable stock.

When the moth-worm has gained a lodgment in the combs, could he have access to the interior, he could attack them in their stronghold.

There are times when it is desirable to know exactly how much honey is on hand. If he could examine the surface of each comb he could determine without difficulty.

In some seasons he has known colonies to so fill their brood-combs with honey, as to allow too little space for breeding, consequently the colony would be small, and all the open cells would not furnish room enough for them to pack themselves away for winter. He sees no remedy for these evils in the common hive.

MOVABLE COMB HIVE

To the Rev. L. L. Langstroth, belongs the credit of introducing to us the hive that will accomplish all these desirable results. Several others have given us hives on the same principle, which effect the same purpose. So many really advantageous points are combined, without interfering with any of the natural wants of the bee, that those of us who appreciate the requisites of a bee hive, and can take advantage of all the facilities offered, can hardly afford to do without some one of these forms, notwithstanding they are covered by a patent.

SOME OF ITS ADVANTAGES.

Each comb, instead of being attached to the top of the hive, is suspended in a frame, and the top is simply laid on loosely. When the bees are dwindling away, and we wish to ascertain the cause, whether queenless, etc., we can take off the top, smoke the bees a little, raise out a comb, and make the necessary examination. Thus, we can also detect the presence of diseased brood. We have only to take a frame from a full hive, and transfer it to the light one and the reverse, to benefit both. To make an artificial swarm, it is only necessary to divide the combs. (See chapter XI.)

When one swarm has issued, we can, seven days after, take out the combs and cut off all queen cells but one, and swarming is stopped for the season. When too much drone comb is constructed, cut it out, and substitute worker comb in its place, fastening it in the same manner as in transferring from the box to the movable comb hives. All suitable comb should be saved for this and similar purposes. The path of the moth-worm in the comb can be traced to his lurking place, and he can be dragged forth to the slaughter without difficulty. The smallest amount of sealed honey can be seen at a glance. The amount of brood that the colony shall raise may be controlled; instead of limiting the area of comb used for that purpose, to a very small space, it may be enlarged to any extent by removing full, and giving empty combs. Notwithstanding the danger of receiving more stings, and the greater expense of construction, there is a class of bee-keepers understanding the value of these conveniences, who will make it pay to use them.

MOVABLE COMB HIVE AS USED BY THE AUTHOR.

I will give a full description and manner of making one, modified by myself from Langstroth's, being much more simple. But he claims that it is not changed sufficiently to be released from his patent.*

DIRECTIONS FOR MAKING.

I make the hive as follows. Get boards twelve and a half inches wide, and one inch in thickness; cut two lengths twenty-one and a half inches, and two, twelve inches. If to be painted, they are planed on both sides, otherwise only inside—these hives having frames on the inside to assist the bees in holding fast, the smooth surface does no harm, and has the advantage of saving the bees the trouble of waxing over the rough places. The two shorter pieces are rabbeted out on the inside upper edge a half inch, to receive the ends of the frames. The whole is now thoroughly nailed together, making a box without top or bottom. The inside is just 12x19½ inches and 12½ deep. At the bottom, in one end, is an entrance three or four inches long, by one-fourth inch deep, also an inch hole half way to the top. The stand and roof are made like those described for the box hive, only longer. The frames for the inside—the point constituting the superiority of the

---

*I am not lawyer enough to decide the point, nor whether the other patents for movable combs are infringements upon his. Therefore I do not wish any one to take the trouble to write to me for an opinion. I am instructed, however, by the owners of several patents, to advise any one disposed to use their hive with no opportunity to purchase right, to use it without hesitation, and when the owner calls on them, if they are ready to pay for an individual right, no harm can be done. It is generally quite agreeable to have their value thus appreciated.

That the reader may have an opportunity of choosing among a variety of hives of this class, I will give the address of several patentees of hives, to whom he may apply for a description:

L. L. Langstroth, Oxford, Butler Co., Ohio.
S. Ide, East Shelby, Orleans Co., N. Y.
T. S. Underhill, Williamsport, Lycoming Co., Pa.
W. C. Harbison, Chenango, Lawrence Co., Pa.
Mr. M. Stillwell, Manlius, Onondaga Co., N. Y.

hive,—are made as follows: First, get out a triangular piece of wood, each side an inch, and eighteen inches long; nail this to one one-fourth inch thick, one inch wide, and twenty and one-fourth long. Each end then projects

Fig. 9.—SIMPLE MOVABLE COMB HIVE.

beyond the triangular piece one and one-eighth inch. Next, get two strips seven-eighths inch wide, by one-fourth inch thick, and eleven inches in length, for the ends, then one for the bottom seven-eighths wide, three-eighths thick, and eighteen inches long, to correspond with the triangular piece at the top. Use small finishing nails, and drive through the ends of the short pieces into the ends of the triangular piece and of the straight piece forming the bottom of the frame. When finished, we have a frame eighteen inches long by ten deep, inside. This will go down into the hive, and leave a half-inch space between the end of the frame and the hive. The strip that is nailed to the triangular one, with projecting ends, rests on the rabbeting and supports it. This is the only part that touches the hive. Eight of these frames will go in a hive that is twelve inches wide, one and one-half inch being the right distance from centre to centre. To keep them from swinging together at the bottom, a stick one-fourth by three-eighths of an inch is put across the middle of the hive three-eighths of an inch from the bottom, with wire braces in this form. Two small mortises, one-fourth inch deep, hold it in place. It may be put in after the hive is together, by bending it a little. Very small annealed wire will do, cut into pieces long enough to reach through, and turn over to the upper side, to hold it firmly The points or angles should be just one and one-half inch apart, and the bottom of the frame should come down between them, within three-eighths of an inch of the piece of wood. If it is desired to have the hive smaller than the above, the places of one or two frames may be filled by a board of the right size; this is better than to vary the size of the hive. It is best to have covers to the boxes all alike, so as to fit all hives.*

Fig. 10.—MOVABLE FRAME.

Fig. 11.—WIRE BRACE TO SUPPORT FRAME.

---

* There may be occasions where it is desirable to have very large hives, such as will hold from 12 to 15 frames. I would suggest that it would be economy for some colonies to have full employment in the hive, in constructing comb and

The top or honey-board easiest made, is a board 21½ inches long, by 14 wide, and three-fourths inch thick, clamped at the ends, with inch holes for passages to the boxes. Such boards are not reliable; notwithstanding the clamps, they will sometimes warp sufficiently to let a bee slip out. Another one that will keep its shape better, is made of several pieces. Two of them are twenty-one and a half inches long, by one and one-half wide; the others eleven inches long, two six inches wide, and two four inches. They are nailed together in this form. The open spaces are for the passages into the boxes which set over them, and are covered with a box that fits the outside of the hive, resting on a piece one-half or three-fourths inch square that is nailed around the hive one inch from the top.

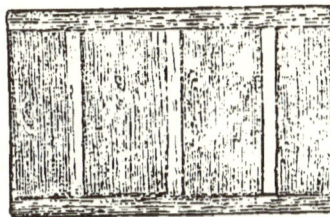

Fig. 12.—HONEY BOARD.

This is the hive that I use principally, and like it rather better than I do Mr. Langstroth's. He has fixtures about his, that must be considered more ornamental than useful, and for which the bees will not perform any extra labor. I am not sure but there are other hives conforming to this principle that would suit me on the whole as well as this. There are some that seem to offer greater conveniences, but cost more. Had I begun with such, I should probably have continued, instead of changing for the more simple one that I have adopted. The convenience of having all

storing winter supplies for those that are deficient. In sections where there is considerable clover and buckwheat, it would be well to have them employed in the hive during the yield of buckwheat, which is of inferior quality, and brings less in market; and get as many combs made and filled as possible, that we may give hives nearly filled, to swarms the next season. We may then put on boxes at once, and there being but little room in the hive, the bees must of necessity store their honey there. This will be the purest quality of clover honey, which would otherwise have been used for the elaboration of wax to fill the hive with combs.

the hives alike is great, and to change all would involve an undesirable expense. I will notice some of the different forms, and the reader may decide for himself, which, under the circumstances, suits him best, and let this be an answer to all who would write me to inquire which I consider the best hive.

With the shape of the hive, and arrangement of the frames, I am satisfied. The depth is all that the comb will sustain when filled with honey, and the greater length of each requires a less number to fill the hive. The bees will store the back end with honey, and rear their brood in the front end, and use nearly every comb for both purposes. This is the rule in properly managed stocks. When winter approaches, there are empty cells in the front end, and honey enough in the other, to last through the cold weather, without obliging the bees to change from one comb to another to obtain it. They have only to move backward as the honey is consumed, on the same principle that they would move upward, in a hive deeper from top to bottom than from front to back. I would not have these frames the longest way up and down for two reasons. Firstly, you could not raise a frame 20 inches in length out of the hive and return it, without hitting the sides occasionally, and arousing the bees. Secondly, there would be too little room on the top for the boxes. When horizontal, there is one-third more room on the top for this purpose. Most of these movable comb hives are nearly square, which shape does not suit me. Some of them have ten or twelve frames, seven or eight inches in depth, by fourteen or fifteen in length. Towards fall, only a part of these, in the middle of the hive, will contain brood; the outside combs are filled throughout with honey. The middle combs contain but little, and the bees begin the winter here. If they are in the cold, and consume the little honey there is in these centre combs, they are quite sure to starve before getting a supply from the outside ones. A winter passage from one comb to the other, is very essential when they are housed, but does not insure their safety in the open air.

Of course any bee-hive can be ornamented according to the fancy of the maker. The plain strip around the top to support the cap, may be a heavy ogee molding, or that cut into dentils would present a tasteful appearance. The cap to cover the boxes may be ornamented in the same way, giving the whole a finished appearance, with but little trouble or expense. When painted let the color be light, and put it on long enough before using, to allow the rank smell of the oil to be lost. To all who use any of the movable comb hives of suitable shape, I would recommend, as a matter of economy, that they make the straw hive also for wintering in the open air, transferring in the beginning of winter. The inside should measure the same every way as the wooden hive, or a very little larger. None of these hives are very good for wintering bees out of doors, without at least a straw mat for the top, similar to the one recommended by Mr. Harbison.

STRAW HIVE FOR WINTERING BEES.

The straw hive that I use is made as follows: First, get out four posts, two inches square, and ten and a half inches long. Then, from an inch board make four strips, two inches wide and twenty-four and a half long, and four of the same width, sixteen and a half inches long. With these, make two flat frames, $16\frac{1}{2} \times 20\frac{1}{2}$ inside, by mortising or halving at the corners. Now, set a post at each corner, and nail through into the end, and the same with the other frame at the other end of the posts, and you have a frame ready to receive the straw. Nail a lath around the middle, inside, another close to the bottom, and one at the top, letting the ends lap on the posts just enough to be held with a nail. Select straight smooth straw,—rye is best—cut it just 10½ inches long, moisten it a little, lay the hive on its side, and put the straw on the lath, till a little more than full, requiring some pressure to make it even with the posts. Some screw or lever is necessary to press it firmly. Laths corresponding with those inside are to be nailed outside, to hold it. Thicker pieces would, perhaps, do better. Pieces 1¼ inches square sawed.in halves, diagonally, would do very well, or pieces turned in a lathe,

Fig. 13.—STRAW HIVE FOR WINTER.

with beads and moldings, cut in two, and the flat side laid next the straw, would improve the appearance.

A strong box to just fit the inside is quite necessary to keep the lath in place while pressing the straw. It is also necessary that the two middle laths are fastened together by a small annealed wire, to keep them from bending from the resistance of the straw when out of the press. The wire should be put on the inner one, and the ends remain projecting through the straw as it is laid on. When the outside piece is nailed on, and before the pressure is relaxed, the wire should be passed around it and twisted to hold it firmly. The ends need nothing of the kind. Make a mat for the top, by framing together vertically, four pieces similar to those used for the hive, rabbet out the end pieces on the lower edge one-fourth inch square, to hold the ends of the three laths to be nailed on at equal distances apart. The straw is filled in and pressed, and pieces nailed over, like the sides just described. Mr. Stilwell has a hive similar to this in principle, (the shape of which I do not like, however,) in which the straw is held by sewing with heavy twine, the manner of holding the straw forming the base of a patent. Whether it is better on the whole, I am unable to say.

For the Leaf or Underhill hive, or any, where the body of combs is separate from the hive itself, a straw box with the top fast to it can be made just large enough to cover the frames. The wooden box for summer is simply to be lifted off, and the other set over.

Glass inserted on one or all sides of a hive, makes it very interesting, but as we now have the movable comb, whereby the interior of the hive may all be brought to light, it is of less consequence than when we were obliged to depend on external observations for all our knowledge of the internal arrangements of the hive.

OBSERVATORY HIVE.

The perfect observatory hive, however, can not fail to be highly interesting to all who feel a curiosity to behold the interior of a bee hive. It can be arranged so readily with one or more frames from a full hive, that all who wish can have one. One comb the size that I use, and a part of a swarm, will exhibit all the phenomena of a full hive. If several frames are used, they may be arranged according to fancy or convenience, one above another, or some above, and others at the ends. A mechanic will construct a special frame to hold them, and a glazed sash to cover each side, giving two inches space between, for the comb. The sash on one side should be movable, that the comb can be changed occasionally, and if more than one is used, that they may all be removed to a regular hive for winter, as such observatory hives are not suitable for cold weather.

CHAPTER IV.

BEE PASTURAGE.

During the warm days of spring, while the winter's snow is melting away, and before the flowers have appeared, the bees seem anxious to be at work. It is then interesting to watch them, and ascertain what they will use as substitutes for pollen and honey.

SUBSTITUTE FOR POLLEN.

At such periods I have seen hundreds engaged upon a heap of saw-dust, gathering the minute particles into pellets on their legs, and seeming quite pleased with the acquisition. Rotten wood, when crumbled into dry powder, is also collected. Flour scattered near the hive is taken up in large quantities. Concerning the utility of flour as a substitute for pollen, I have now had considerable experience. Yet much depends on the locality, number of the bees, and quantity of snow. Where there are but few bees, and little snow, the early flowers appear so soon after the bees begin to fly, that flour is of but little advantage. But when the number of bees greatly exceeds the supply, the flour should be given during pleasant days, to promote early breeding, and establish habits of industry, as well as to prevent marauding, which is very important.

To feed it advantageously, make a floor a few feet square, with a curb around, three or four inches high, to prevent waste. When practicable, feed rye, ground very fine, and unbolted. The bees seem to like to work out the flour and fine particles from among the bran, better than to work in clear flour. Yet the latter will do when the former is not to be obtained, but should be mixed with cut straw or saw-dust. The bran left by the bees may be fed to other stock. When the flowers yield pollen in sufficient quantities, they will no longer take the flour. It should be remembered that flour feed is only advantageous in the earliest part of the season. Unless it can be given then, it is useless to take the trouble.

SUBSTITUTE FOR HONEY.

A substitute for a small quantity of honey is found in the sap of a few kinds of trees. A syrup made from sugar is a very good substitute for honey.

MANNER OF PACKING POLLEN.

The particular manner of packing pollen has been satisfactorily witnessed by but very few persons, as the operation is mostly performed on the wing, thereby preventing a fair chance for minutely inspecting it. When collecting pollen only, they light upon the flowers, and pass rapidly over the stamens, detaching a portion of the dust, which lodges on most parts of them, and is brushed together, and packed into pellets when they are again on the wing. While the bees are gathering flour, the process is more readily seen.

The Italians may often be seen appropriating old bits of comb that have been squeezed together, and propolis from old boards of broken hives. They merely bite off little particles, and pack them on their thighs, before they rise on the wing. As soon as a load is obtained, they immediately return to the hive, each bee bringing several loads in a day. Honey, as it is collected, is deposited in the abdomen, and kept out of sight until stored in the hive.

The time that bees commence their labors in the spring does not by any means govern the time of swarming; this depends upon the weather through April and May.

FLOWERS THAT YIELD THE FIRST POLLEN.

The first material gathered from flowers is pollen. Common or Candle Alder, (*Alnus serrulata*) and Skunk Cabbage, (*Symplocarpus foetidus*,) yield the first supply. In this latitude (42°) their time of flowering varies from March 10th to April 20th. The amount of pollen they afford is also variable. Cold freezing weather frequently destroys a great portion of the flowers after they are out. The staminate flowers of the alder are nearly perfected the previous season, and a few warm days in spring will develop them before any leaves appear. When the weather continues fine, great quantities of pollen are secured. Our swamps produce several kinds of willows (*Salix*) that put out their blossoms very irregularly. Some of these bushes are a month earlier than others, and some of the buds on the same bush are a week or two later than the rest. These also afford only pollen, but are a much more sure dependence than the alder; a turn of cold weather can not at any time destroy more than a small proportion of the flowers. The Aspen, (*Populus tremuloides*) which comes next, is not a particular favorite with the bees, as but few, comparatively, visit it. It is followed very soon by an abundance of the Red Maple, (*Acer rubrum*,) that suits them better, but this, like some others, is often lost by freezing.

FIRST HONEY.

The first honey of any account is obtained from the Golden Willow, (*Salix vitellina*); which is seldom injured by frost. Gooseberries, currants, cherries, pear and peach trees contribute a share of both honey and pollen. Sugar Maple, (*Acer saccharinum*) throws out its ten thousand beautiful silken tassels with a bounteous yield of tempting nectar. Strawberries modestly open their petals in invitation, but like " obscure virtues," are often neglected for the more conspicuous Dandelion, and the showy and fragrant blossoms of the Apple, which now open their stores, and offer to the bees a real harvest.

FRUIT FLOWERS IMPORTANT.

In good weather, a gain of 20 pounds is sometimes added to the hives during the period of apple blossoms. But we are seldom fortunate enough to have continuous good weather, as it is often rainy, cloudy, cool or windy, all of which conditions are very detrimental. A frost will sometimes destroy all, and the gain of our bees is reversed, that is, their stores are lighter at the end than at the beginning of this season of flowers. Yet this season often decides the prosperity of the bees for the summer. If there is good weather now, we expect our first swarms about June 1st; if not, no subsequent yield of honey will make up the deficiency.

We now have a time of several days, from 10 to 14, in which there are but few flowers. If our hives are poorly supplied when this scarcity occurs, it will so disarrange their plans for swarming, that no preparations are again made much before July, and sometimes not at all. In sections where the wild cherry, (*Prunus serotina*) abounds, these flowers will appear, and fill the period of scarcity which this section annually presents. The Locust, (*Robinia Pseudacacia*), blossoms at this time, and where it is sufficiently abundant, is valuable as bee food, while it is also well worthy of cultivation for timber.

RED RASPBERRY A FAVORITE.

The Red Raspberry, (*Rubus strigosus*) now presents the stamens as the most conspicuous part of the flower, soliciting the attention of the bee, by pouring out the bounteous libations so highly prized by our industrous insect. For several weeks they are allowed to partake of this exquisite beverage; it is secreted at all hours, and in all kinds of weather. When the morning is warm, we often hear their cheerful humming among the leaves and flowers of this shrub, before the sun appears above the horizon. The gentle shower, sufficient to induce man to seek shelter, is often unheeded by the bee when luxuriating among these flowers; even white clover, important as it is in furnishing the greatest part of the stores, would be neglected at this season, if the raspberry only yielded a full supply. Clover begins to blossom with the raspberry, and continues longer.

HONEY FROM RED CLOVER.

Red clover probably secretes as much honey as the white, but the tube of the corolla being longer, common bees appear to be unable to reach it. I have seen a few at work upon it, but it appeared to be slow business. The Italians work on it sometimes, apparently out of choice.

Sorrel, (*Rumex acetosella*), the pest of many farmers, is brought under contribution by the bees, and furnishes pollen in any quantity. Morning is the only part of the day appropriated to its collection.

CATNIP ONE OF THE BEST HONEY YIELDING PLANTS.

Catnip (*Nepeta Cataria*), Motherwort, (*Leonurus Cardiaca*), and Hoarhound (*Marrubium vulgare*,) put forth their flowers about the middle of June, rich in sweetness, and like the raspberry, the bees visit them at all hours, and in nearly all kinds of weather. They remain in bloom from four to six weeks; in a few instances I have known the catnip to last twelve, yielding honey during the whole time. If there is any plant that I would cultivate especially for honey, it would be catnip. I find nothing to surpass it. Borage has been recommended as yielding abundantly, and worthy of cultivation, but the profusion of flowers produced by the catnip, seems to excel it. The Alsike or Swedish white clover has also much to recommend it. The plant being valuable for soiling cattle, or for hay, would be a desirable acquisition to the bee-keeping farmer, as well as to others on whose land it will thrive. It does not do well on sandy soil, with me.

Ox-eye Daisy (*Leucanthemum vulgare*), a beautiful flower in pasture and meadow, and worth but little in either, also contains some honey. The flower is compound, and each little floret secretes so minute a quantity that the task of obtaining a load is very tedious. It is only visited when the more copiously honey-yielding flowers are scarce. The Toad-flax or Snap-dragon, (*Linaria vulgaris*), with its disagreeable odor, troubling the farmer with its vile presence, is made to bestow the only good thing about it, except its beauty, upon our insect. The flower is large and tubular, and to reach the honey the bee must enter it. To see the bee almost disappear within the folds of the corolla, one would think it was about being swallowed, but it soon emerges, covered with dust, unharmed, from the yellow prison. This is not brushed into pellets on its legs, like the pollen from some other flowers, and some adheres to its back, between the wings, which it is apparently unable to remove, as it often remains there for months. Bush Honey-suckle, (*Diervilla trifida*), is another particular favorite.

4*

SINGULAR FATALITY ATTENDANT ON SILKWEED.

Silkweed, (*Asclepias Cornuti*), is another honey-yielding perennial, but a singular fatality befalls many bees while gathering honey from it, that I never have seen noticed. I have observed during the period this plant was in bloom, that a number of the bees belonging to hives not full, were unable to ascend the sides to the comb; there would be some times thirty or more at the bottom in the morning. On searching for the cause, I found from one to ten thin yellow scales, of a long pear-shape, and about the twentieth part of an inch long, attached to their feet: At the small end was a black thread-like substance, from a sixteenth to an eighth of an inch in length; on this stem was a glutinous matter, that firmly adhered to each foot or claw of the bee, preventing it from climbing the sides of the hive. I also found this appendage attached to bees clustered outside of full hives, but it appeared to be no inconvenience to them. Among the scales of wax, and waste matter that accumulates about the swarms to some extent, I found a great many of these scales which the bees had worked from their feet. The question then arose, were these scales a foreign substance, accidentally entangled in their claws, or was it a natural formation? It was soon decided. From the number of bees carrying it, I was satisfied that if it were the product of any flower, it belonged to a species somewhat abundant. I made a close examination of all such as were then in bloom. I found the flowers of the Silk-weed or Milk-weed, sometimes holding a dead bee by the foot, secured by this appendage. Both sepals and petals of this flower are turned backward towards the stem, forming five acute angles or notches, just the trap for a bee with this attachment. When at work, they are very liable to slip a foot into one of these notches; the flower being thick and firm, holds it fast, and pulling only draws it deeper in the wedge-like cavity. The appendage which causes so much trouble to the bees, is the pollen of the Silk-weed, which in all the species has a singular form. Instead of being, as is the case in most flowers, a fine dust, the pollen grains are stuck together in little waxy masses or scales, and these are joined together in pairs by the thread-like appendage above noticed. These masses are, in the flower, each lodged in a little pouch with only the attachment exposed, and were it not for the agency of bees and other insects, the pollen would not be dislodged from these pouches and brought in contact with the pistil of the flower. When I point out a loss among bees, I would like to give a remedy, but here I am unable to do so. I am not sure but honey enough is obtained by such bees as escape, to counterbalance the loss.

Whitewood, (*Liriodendron Tulipifera*), yields something eagerly sought for by the bees, but whether honey or pollen, or both, I have never ascertained. Mr. Harbison asserts it to be honey. I have never examined the flowers. It is very scarce in Montgomery and Greene Counties. Mr. Langstroth speaks of it as "one of the greatest honey-producing trees in the world. As its blossoms expand in succession, new swarms will sometimes fill their hives from this source alone."

BASSWOOD VERY IMPORTANT.

Basswood, (*Tilia Americana*), is abundant in some places, and yields honey clear and transparent as water, of a delicious flavor, with a perceptible, yet not unpleasant taste of mint. During the time this tree is in bloom, a period of two or three weeks, in many sections, astonishing quantities are obtained when the weather is favorable. It is less likely to be cut off by bad weather, than other blossoms. A person once assured me that he had known ten pounds of honey collected in a day, while this was in flower, by one swarm. I have seen a statement by a writer in Wisconsin, that "hives have increased in weight one hundred pounds while this tree was in bloom." I think these statements are quite as large as can be credited. I have no comparable experience. I have weighed hives during the seasons of apple blossoms, buckwheat and clover, the best source of honey wherever I have kept bees, and three and one-half pounds is the greatest yield I ever found in one day. As a shade tree, Basswood, or as sometimes called, Linden, ranks with the finest. It is hardy, and bears transplanting better than most kinds. This stately tree with its graceful clusters of fragrant flowers, adorns village or country grounds, while the soft music of the industrious bee, among the branches, is attractive to the dullest ear. The honey resources of the country might be greatly increased by planting such trees.

Sumach, (*Rhus glabra*), is rich in its quality and yield of honey. The shrubs coming into bloom in succession, the supply is protracted beyond the duration of one set of blossoms. Mustard, (*Sinapis nigra*), is also a great favorite. Its cultivation is remunerative for its seed alone, and when we add the advantage that it is to the bees, there seems to be a sufficient inducement to cultivate it.

I have now mentioned most of the honey-producing trees and plants, which bloom before the middle of July. The course of these flowers is termed the first yield. In sections where there are no crops of buckwheat, it constitutes the only full one. Other flowers continue to bloom until cold weather. Where white clover is abundant, and the fields are used for pasture, it will continue to throw out fresh flowers, sometimes, throughout the entire summer, yet the bees consume about all they collect, in rearing their brood, etc. Thus, it appears, that in some sections the bees have only about six or eight weeks in which to provide for winter.

HONEY DEW.

Honey dew is said to be a source whence large collections are made in some places. When or where it appears or disappears, is more than I can tell, from my own experience. Twelve years ago, I expressed what was taken for doubt of the existence of any such substance. To enlighten me on the subject, and give ocular demonstration, some of my friends, living where it was found, have kindly sent me specimens —leaves covered with it—for my inspection. It appeared and tasted as if some saccharine substance somewhat diluted, had been spread evenly over the upper side of the leaf, and the watery particles had evaporated. I have seen descriptions of it as found somewhere, well towards the Golden State, that exceeded any thing I ever heard of in the Eastern States. It was described as covering leaves and branches in such quantities as to bend them down with the excessive weight. The quantity was so great as to induce an effort to collect it by hand. The question as to its origin has been pretty thoroughly discussed without arriving at any particularly clear conclusion. It is generally attributed to the Aphis or Plant Louse. It will be seen that this theory of its origin will account for only a part of the phenomena. Some years ago, in the month of August, I noticed on passing under some willow trees, (*Salix Vitellina*), that the grass and stones were covered with a wet or shining substance. I found that nearly all the smallest branches were covered with a species of large aphis, apparently engaged in sucking the juices, and occasionally discharging a minute drop of a transparent liquid. I *guessed* this might be honey dew. I visited the place again after sunrise, to see if there were any bees collecting it. I found them in hundreds, together with ants, hornets, and wasps. Some were on the branches with the aphis, others on the leaves, and some on the grass and stones. This liquid, ejected by the aphis when sucking the juices of tender leaves or branches, and received by ants that are usually in attendance, is probably the honey dew of many writers. Ants, instead of bees, generally collect it. These insects have been very appropriately termed "ants' cows," as they are regarded by them with the most tender care and solicitude. In July or August, when the majority of the leaves of the apple trees are matured, there are often a few sprouts or suckers about the lower part of the trunk, that continue growing and putting out fresh leaves. On the under side of these, you will find this insect by hundreds, of all sizes, from those just hatched to the perfect aphis. All appear to be engaged in sucking the bitter juice from the tender leaf and stalk. The ants are among them by scores. The careless observer often accuses them of doing the injury instead of the aphis. Occasionally there will issue from the abdomen of the aphis a small transparent globule, which the ant is ready to receive. When a load is obtained, it descends to the nest. Many other kinds of trees and plants are used by the ants as "cow pasture," and most kinds of ants are engaged in this dairy business. Would the bees attend the aphis for this secretion if the ants left any to be gathered? Or, if there were no ants or bees, would this secretion be discharged and falling on the leaves below them, be honey dew? If they were situated on some lofty trees, and it lodged on the leaves of small bushes nearer the earth, it would be considered such by some.

UNUSUAL SECRETION.

I once discovered bees collecting a secretion unconnected with flowers, but which was not honey dew, as it has been described. I was passing a bush of Witch-hazel, (*Hamamælis Virginiana*), and my attention was arrested by an unusual humming of bees. At first I supposed that a swarm was about me, yet it was late in the season, July

25th. On close inspection, I found numerous warty excrescences upon the bush, of the size and shape of a hickory nut. These proved to be only shells, the inside being lined with thousands of minute insects, a species of aphis. These appeared to be sucking the juices, and discharging a clear transparent fluid. Near the stem was an orifice about an eighth of an inch in diameter, out of which this liquid exuded gradually. So eager were the bees for this secretion, that several of them crowded around one orifice at a time, each endeavoring to thrust the other away. This occurred several years ago, and I have never been able to find any thing like it since, neither have I learned whether it is common in other sections.

Within a few years past, a species of aphis has appeared on the grain in many sections, covering the straw in myriads, sucking the juices and secreting at the time a saccharine substance, which is collected by the bees. Correspondents from some of the Western States, particularly Wisconsin, write that the bees gather large quantities of this, and that as winter food it proves unhealthy, causing dysentery, etc. I have received numerous applications for a remedy, but as I have not had the least experience, I cannot advise. This secretion being more animal than vegetable, is an unnatural aliment for the bee, and as might be expected, is unhealthy. According to the prediction of Dr. Fitch, this race of insects will soon disappear, and our bee-keeping friends may expect better times. I remember hearing it predicted when I was a boy, that a certain winter would " be a bad one for bees," because they were seen obtaining honey dew from hickory leaves. The question arises, Was the effect of an unnatural substance taken by the bees forty years ago, similar to that produced by the secretions from the aphis in later years? All this does not explain the origin of honey dew, unless we admit two or more sources. Honey dew is found in the open field where no tree is standing above to shower it upon leaves below. It is found on leaves, having no traces of the aphis near them. How did it get there? Did the leaves secrete it? I am not yet ready to admit this. If leaves produce it, why is it not found in this section?

In passing I have not mentioned garden flowers, because the amount obtained from them, especially ornamental flowers, is inconsiderable, compared to that from forest and field. It is true that the Hollyhock, (*Althea rosea*), Mallows, (*Malva rotundifolia*), Mignonette, (*Reseda odorata*), and many others yield honey, but of small account. A person who expects to have his hives filled from such a source, will be very likely to be disappointed, unless his number of stocks is very limited.

We will now notice the flowers that appear after the middle of July. The Button-ball, (*Cephalanthus ocidentalis*) is much frequented for honey. Also our vines—melons, cucumbers, squashes and pumpkins. The latter are visited only in the morning, and honey is the only thing obtained. Nothwithstanding the bee is covered with farina, it is not kneaded into pellets on its legs. I have seen it stated that bees get pollen early in the morning, instead of honey. It is not best to always take our word, about such matters, but examine for yourselves. Take a look some warm morning, when the pumpkins are in bloom, and see whether it is honey or pollen of which they are in quest.

Under some circumstances, clover will continue to bloom through this part of the season, and a few other flowers also, but I find by weighing, a loss from one to six pounds between July 20th and August 10th, at which time Buckwheat usually begins to yield honey, which generally proves a second harvest.

BUCKWHEAT HONEY.

In several counties in this State, so little of this grain is raised, that the honey can not be found in the hive or boxes. But in many places it is the main dependence, the bees seldom getting more than a winter supply from the early flowers. This honey is considered by many to be of inferior quality. Its color, when separated from the comb, resembles molasses of medium shade. The taste is more pungent than that of clover honey; it is particularly prized on that account by some, and disliked by others for the same reason. When swarms issuing as late as July 15th, commence on buckwheat, they will sometimes contain not more than five pounds of stores, and yet make good stocks for winter, whereas without this yield, they might not live through October. This crop fails about once in ten years. I have known a swarm to gain sixteen pounds in one week, and construct comb to store it in at the same time. I once had a swarm issue August 18th, that obtained 30 pounds in about eighteen days. But such buckwheat swarms, in ordinary seasons, seldom get over 15 pounds. The buckwheat flowers last from three to five weeks. The time of sowing varies in different sections, from June 10th to July 20th. Farmers wish to give it just time to ripen before frost, as the yield of grain is considered better, but as the time of frost is a matter of uncertainty, some sow several days earlier than others. Whenever an abundant crop of this grain is realized, a proportionable quantity of honey is obtained.*

DO BEES INJURE THE GRAIN?

Many people contend that bees are an injury to this crop, by taking away the substance that would be formed

---

* A friend informs me that, in 1863, the bees in some parts of Albany Co., N. Y., refused to swarm before buckwheat blossomed, and that between the 1st and 10th of August he had one hundred swarms. Many of them stored abundance for winter, and gained considerable surplus; in some instances, 28 lbs.

into grain. The best reasons that I have obtained for such an opinion, are these: "I believe it, and have thought so for a long time." "It is reasonable, that if a portion of the plant is taken away by the bees, there must be less material left for the formation of seed, etc." Most of us have learned that a person's opinion is not the strongest kind of proof. Are the above reasons satisfactory? How are the facts? The flowers open, and honey is secreted. If the bee does not lick it up, it dries up and is wasted. Now, what is the difference to the plant, whether the honey is lost in this way, or is collected by the bees? If there is any difference, the advantage appears to be in favor of collection by the bees, for the reason that it thus answers an important end in the economy of nature, consistent with her provisions in ten thousand different ways of adapting means to ends. Most breeders of domestic animals are aware of the degeneration induced by in-and-in breeding, and that a change of breed is necessary for improvement, etc. Vegetable physiology seems to indicate a similar necessity among plants. The stamens and pistils of flowers answer for the two sexes in animals. The pistil is connected with the ovaries, and the stamens furnish the pollen that must come in contact with the pistil; in other words, it must be impregnated by this dust from the stamens, or no fruit will be produced. Now, if it be necessary to change the breed, or essential that the pollen produced by the stamens of one flower shall fertilize the pistils of another, to prevent barrenness, what could we contrive better than the arrangement already made by Him who knew the necessity, and planned it accordingly! And it works so admirably that we can hardly avoid the conclusion that this was an important part of the design in creating bees. Their food consists of honey and pollen; each flower secretes but a little, just enough to attract the bee, for nothing like a full load is obtained from one;

were it otherwise, the end in view would not be answered. A hundred or more flowers are often visited in one excursion, and the pollen obtained from the first may fertilize many others previous to the return of the bee to the hive. By such a cross-fertilizing, a field of buckwheat may be kept in health and vigor in its future productions. A field of wheat produces long slender stalks that bend to the breeze, and one ear is made to bestow its pollen on an ear several feet distant, thereby effecting just what bees do for buckwheat. Corn, from its manner of growth, the upright stalks bearing the stamens some feet above the pistils on the ears below, seems to need no agency of bees; the superabundant pollen from the tassel is wafted by the wind several rods from the stalk that produces it, and there does its work of fertilizing the distant ear, as is proved by the mixing of different varieties at some distance.

BEES NECESSARY TO INSURE A CROP.

But how is it with the vines trailing on the ground, a part of the flowers producing stamens, the others pistils? It is absolutely essential to produce fruit, that pollen from the staminate flowers shall be introduced into the pistillate ones; if this fails to occur, the germ will wither and die. In the bee we have an agent ready for the purpose; both staminate and pistillate flowers are visited promiscuously by it, the pollen, not being kneaded into pellets, (particularly that from pumpkins,) adheres to every part of the body, rendering it next to impossible for the bee to enter a pistillate flower without leaving a portion of the fertilizing dust in its proper place. Hence it is reasonably inferred by many, that if it were not for this agent among our vines, the uncertainty of a crop from non-fertilization, would render their cultivation a useless task. When the aphis is located on the stalk or leaf of a plant, it is furnished with means to pierce the surface, and extract the juices essential to its formation, thereby preventing a vigorous growth and full development. This idea is too apt to be associated with the bee when it visits the flower, as if it were armed with a spear, to pierce bark or stem, and rob it of its nourishment. An examination of the structure of the bee will show us that this cannot be the case. Its slender, brush-like tongue, folded closely under its neck, and seldom seen except when in use, is not fitted to pierce the most delicate substance; all that it can be used for is to sweep or lick up the nectar as it exudes from the flower; this is secreted for no other purpose, it would seem, than to attract the bee. The most delicate petal receives no injury while the bee is using the instrument nature has provided for obtaining the sweets. During one excursion the bee seldom visits more than a single species of flower; were it otherwise, and all kinds were visited promiscuously, the fertilizing of one species with the pollen from another, would be quite likely to produce some hybrids among plants. Writers, when noticing this peculiarity of instinct, cannot be content, but must add other marvels. They follow this trait still farther, and make the bee store every kind by itself in the hive.

TWO KINDS OF POLLEN STORED IN ONE CELL.

With regard to honey it is not easy to ascertain; but pollen is of different colors, generally yellow, but sometimes pale-green, and reddish or dark-brown. I think a little patient inspection will satisfy any one that two kinds are sometimes packed in one cell. I will admit that two colors *are* seldom found thus, but it is sometimes the case. I have found it thus, and proved this assertion worthless.

NO TEST OF THE PRESENCE OF THE QUEEN.

It is asserted that "if a hive loses its queen, no pollen is collected." Also, "that such quantities are sometimes collected, and so many cells are filled, that too little room is left for brood, and the stock rapidly dwindles in consequence." The first of these assertions has been offered as a test to determine the presence of a queen. My bees have such a habit of doing wrong, that it is no test whatever. I think I can explain the mystery of a stock containing an unusual quantity of bee-bread with the honey, and show that instead of this being the *cause* of a scarcity of bees, it is the *effect*. Stocks and sometimes swarms lose their queen in the swarming season—(see particulars in chap. x,) when, instead of remaining idle, they collect the usual quantity of pollen and honey. There being no larvæ to consume the pollen, the consequence is, more than half the breeding cells will contain it; they will be packed about two-thirds full, and finished out with honey. I have known large families to be left under such circumstances, and about all the cells in the hive were thus occupied. Whereas, in a stock containing a queen, and rearing brood, a portion of the combs will be used for this purpose until the flowers fail, when such comb will be found empty. In order to ascertain whether this extra quantity of the bee-bread was so *very* detrimental, I have introduced into such hive in the fall, a family with a queen, wintered them in it, and watched their prosperity another year, and never found them unprofitable on that account. I am so well satisfied of this that whenever I now have a hive in such a situation, I make it a rule to introduce a colony with queen.

It is generally calculated that when medium sized hives are full, about seven-eighths of the cells are of the proper size for raising workers; the remainder, except a few designed for queens, are of the size for drones.

BEE BREAD SELDOM PACKED IN DRONE CELLS.

Bee bread is generally exclusively packed in the worker cells. I might as well remark here, that when taking combs from a hive filled with honey, if such pieces were selected as contained drone cells, there would be but little risk of finding bee-bread; the outside sheets, and the upper corners of the others are next best. The sheets of comb used principally for raising workers, and the cells adjoining those used for breeding, for an inch or two in width, are nearly all packed with pollen, and much of it will remain when the breeding season is past. Smaller portions are found in the worker cells in nearly all parts of the hive; even the boxes will sometimes contain a little.

MANNER OF DISCHARGING POLLEN.

In a glass hive the bees may be seen depositing their loads of pollen. The legs holding the pellets are thrust into the cell, and a motion like rubbing them together is made for half a minute, when they are withdrawn, and the two little loaves of bread may be seen at the bottom. This bee appears to take no farther care about them, but another will soon come along, enter the cell head first, and pack it close. The cell is filled about two-thirds of its length in this way, and when sealed over, a little honey is used to fill it out. To witness the operation of depositing honey, a glass hive or box is requisite, as the edges of the combs will be attached to the glass.

DISCHARGING HONEY.

When honey is abundant, most of these half cells next the glass will contain some. The bee goes to the bottom of the cell, deposits a particle of honey, and brushes it into the corners or angles with its tongue, carefully excluding all the air. As it is filled, that next the sides of the cells is kept in advance of the centre. This is just as a philosopher would say it should be done. If it were filled at once, and no care taken to attach it to the sides, the external air would not keep it in place, as it now does, effectually, when the cell is of ordinary length. When the cell is about one-fourth of an inch deep, they often commence filling it, and as it is lengthened, they continue to add honey, keeping it within an eighth of an inch of the ends, it is never quite full, till nearly sealed over, and often not then. In worker cells the sealing seldom touches the honey. But in drone cells the case is different. The honey on the end touches the sealing about half way up. It is kept in the same concave shape while being filled, but being in a larger cell, the atmospheric pressure is less effectual in keeping it in its place; consequently, when they commence sealing these cells, they begin on the lower side, and finish at the top. When storing honey in boxes, cells of this size are usually much longer, in which case they are crooked, the ends turning upward, sometimes half an inch or more. This, of course, will prevent the honey from running, but if the box is taken off, and turned over before such cells are sealed, they are very sure to lose much of their contents. The drone cells of ordinary length, in the breeding department, will hold the honey well enough as long as they remain horizontal, but turn the hive on its side, and bring the open end downward, in hot weather, or break out a piece and hold it in that position, the air will not keep the honey in place, but will do so in the worker cells.

SOME CELLS CONTAINING HONEY FOR DAILY USE.

.I never examined a hive, fully supplied with bees and honey, in winter or summer, but it had a number of unsealed cells containing honey, as well as pollen, unless it was destitute of a queen. They will always have some cells open for daily use, even if they have stored a large quantity in boxes, and are so crowded for room as to store honey outside, or under the bottom boards.

COMBS CONSTRUCTED AS NEEDED.

Young swarms seem unwilling to construct combs faster than needed for use. This would appear, at first thought, to be a lack of economy. When no honey is obtained, and there is nothing to do, it would seem to be well to get ready for a yield, but this is not their way of doing business. Whether they can not spare the honey already collected to elaborate the wax, or whether they find it more difficult to keep the worms from a large quantity of comb, I shall not presume to decide. If honey is abundant, large swarms, when first located, will extend their combs from top to bottom in a little more than two weeks, but such hive is not yet full. Some sheets of comb may contain honey throughout their entire length, and not a cell be sealed over, but the bees generally find time to finish up to within a few inches of the lower end as they proceed. Whenever unfinished cells contain honey, it will generally be removed soon after the flowers fail, and used before that which is sealed, and the cells will remain empty till another year.

BEST SEASON FOR HONEY.

The inquiry is often made, "Which is best for bees, a wet or dry season?" I have studied this point very closely, and have found that a medium between the two extremes produces the most honey. When farmers begin to express fears of a drought, then is the time, if in the season of flowers, that most honey is usually obtained, but if dry weather is much protracted, the quantity is greatly diminished. Of the two extremes, a very wet season is perhaps the worst.

HOW MANY STOCKS MAY BE KEPT.

"What number of stocks can be kept in one place?" is a question so often asked that it indicates an unusual interest in the subject. I shall differ more in opinion here, with some of our best authors, than on most other points. Mr. Langstroth expresses himself very confidently that over-stocking has never happened in this country, and that there is no prospect of it. He gives us, on the authority of Mr. Wagner, the number of stocks to the square mile in many sections of Europe. I will give one or two items. "In the Kingdom of Hanover, 141 stocks, are estimated to the square mile." "In the Province of Atica, in Greece, containing 45 square miles, 20,000 hives are kept." "A Province in Holland contains 2000 colonies per square mile."

The honey yielded from the flowers in this section, (Montgomery Co., N. Y.,) in 1863 would have supported but a small part of that number, through the season. As it was, we had too few and too many. Let me explain. From about the 15th to the 30th of June, clover yielded honey, and the bees seemed to improve the time, industriously storing the usual quantity. I presume that during *this* period, thrice the number would have done equally well. Those who have recommended keeping such large numbers must have had such a yield as this in view. In Europe, where so many are reported to be kept, it must be thus throughout the season. But with us, after the latter date, but few plants produced honey; even Basswood seemed to yield but little. A few plants, such as catnip, motherwort, and silkweed, furnished enough to have kept a half dozen colonies in thriving condition, but when this amount was divided among hundreds, there was not enough to keep all alive. When buckwheat blossomed, there was perhaps enough for half a dozen hives to the square mile, and this number might have shown results equal to those in Albany Co.,—50 pounds to a colony—as the flowers appeared to yield abundantly, each hive obtaining five or six pounds. Is it not evident that we were overstocked after July 1st? The summer of '64 gave a bounteous yield of honey until July 10th, when the supply was diminished, probably by drought. I say *probably*, because we can not always tell to what cause to attribute the greater or less abundance of the supply. Before said date, any number of colonies, apparently, would have done well, but since that time, one-tenth of the actual number kept would have collected the whole yield. Yet it was profitable to keep about fifteen to the square mile. It will always be impossible to know exactly how many can be kept; some seasons produce bounteously, others, a partial supply, and some almost none at all. As it is difficult to tell beforehand what to expect, it is well to exercise some caution. Whoever begins with excessive numbers, must expect sometime to be overtaken with serious disaster. The sight of a hundred or two colonies, actually starving in December, is rather unpleasant to a sensitive mind. One must lay up a store of fortitude, in prosperous times, to last him through such seasons of discouragement.

It is an advantage to keep as large a number as will possibly do well in one yard. They may be taken care of with much less proportional expense. It would not do to hire a man to take charge of every eight or ten hives, although the average profit of the few would be much greater than with a large number. One man can take care of 200 stocks, especially if he uses the movable comb hive, and the reduction in the expense would more than balance the larger profits from the smaller apiaries. I would not advise keeping very large apiaries, until warranted by experience in their care. Also the resources of a country should be gradually tested. A honey-producing country may be like a grazing region. One field may pasture ten times as many cattle as another, and the same difference may be true of pasturage for bees.

PRINCIPAL SOURCES OF HONEY.

There are three principal sources of honey, viz: clover, buckwheat and basswood. Clover is the only *universal* dependence, as that is found almost everywhere, in greater or less profusion. Buckwheat is the main source in some places. Basswood is of brief duration, but comes in very opportunely where it abounds, just as clover begins to fail, and before buckwheat appears. Where all these are abundant, there is the true Eldorado of the apiarian. Yet to find a place where there is a great plenty of both clover and buckwheat is very difficult. I have failed after a long and patient search. I find clover, without buckwheat, in satisfactory abundance. But when I begin to find buckwheat, clover correspondingly diminishes. Where buckwheat is a universal crop, but little of the surplus honey is clover, as in the counties of Greene and most of Albany. The question is asked: "What section of country is best for keeping bees?" It is difficult to answer. In clover regions the superior quality and enormous crops of honey in some seasons will give very desirable results, but when an occasional failure occurs, it is disastrous in the extreme. In buckwheat sections, there is never a great yield of clover, but seldom a failure in buckwheat honey.

If the first yield fails, the last usually supplies the deficiency, and all strong colonies will generally have sufficient winter stores. I have now been speaking of large apiaries. A section can hardly be found where man can live, where a *few* stocks would not thrive, even if no dependence could be placed on the prominent sources just mentioned. There will be some honey-yielding flowers in nearly all places.

DISTANCE THAT A BEE WILL GO FOR HONEY.

Another question of interest is concerning the distance a bee will travel in search of honey. It is evident that it will be farther than for purposes of plunder. I have heard of their being found seven miles from home. It was said to be ascertained by sprinkling flour on them as they left the hive in the morning, and discovering the bees thus marked at that distance from home. When we consider the chances of finding a bee, even one mile from the hive, this appears rather dubious; and likewise, pollen, the color of flour, might deceive a casual observer, or one who had a case to sustain. It is difficult to prove that they go three miles. I think from present evidence that they do not go farther. The queens and drones, situated that distance apart do sometimes meet, as is proved by black queens producing hybrids, but whether one travels the whole distance, or they meet each other half way, is not certain. I have my yards from two and a half to five miles apart. The largest apiaries should be separated, at least, four miles.

CHAPTER V.

THE APIARY

LOCATION.

One *important* consideration, in the location of an apiary, is in regard to convenience for watching in the swarming season. If much trouble has to be taken, it is too often neglected. Unless the apiary is large, watching need not occupy one's whole time, but it may be done in connection with some other employment, and it is desirable to have the hives located with reference to this. Although the movable combs may be used, and each stock divided as it is filled, and no swarms expected, yet one will occasionally issue, making some attention necessary. If possible, the hives should stand where the wind will have but little effect, especially from the north west. If no hills or buildings offer a protection, a close high board fence should be put up for the purpose. The saving of bees will pay the expense. During the first spring months, the stocks contain fewer bees than at any other season. It is then that a large family is important, to keep the brood warm. One bee is of more consequence then than a dozen in midsummer. When the hive stands in a bleak place, the bees returning with heavy loads, in a high wind, are frequently unable to strike the hive, are blown to the ground, and become chilled and die. A chilly south wind is equally fatal, but not so frequent. When protected from winds, the hives may front as you choose; east or south is generally preferred. A location near ponds, lakes, large streams, etc., involves some loss. Hard winds fatigue the bees when on the wing, often causing them to alight in the water, whence it is impossible to rise again until wafted ashore, and then, unless in very warm weather, they are so chilled as to be past recovery. I do not mention this to discourage any one from keeping them, when so situated, because some must keep them thus, or not at all. Although we can not miss a few lost from each stock, it is nevertheless a loss as far as it goes.

Whatever location is chosen, it should be decided upon as early in the spring as possible, because when the chilling winds of winter have ceased for a day, and the sun, unobstructed, is sending his first warm rays upon the frozen earth, the bees that have been inactive for months, feel the cheering influence, and come forth to enjoy the balmy air.

LOCATION MARKED.

As they come from their door, they pause a moment, as if to rub their eyes, which have been so long obscured in darkness. They rise on the wing, but instead of leaving in a direct line, immediately turn their heads towards the entrance of their tenement, describing a circle of a few inches at first, but larger as they recede, until an area of several rods has been viewed and marked.

SHOULD NOT BE MOVED.

After a few excursions, and surrounding objects have become familiar, this precaution is not taken, and they leave in a direct line for their destination, and return by their way-marks without difficulty. Man, with his reason, is guided in the same manner. There are a great many people who suppose the bee knows its hive by a kind of instinct, or is attracted towards it, like the steel to the magnet. At least they act as if they thought so, as they often move their bees a few feet or rods, after the location is thus marked, and what is the consequence? The stocks are materially injured, and sometimes entirely ruined by loss of bees. Let us notice the cause. As I have stated, the bees have marked the location. They leave the hive without any precaution, as surrounding objects are familiar. They return to their old stand and find no home. If there is more than one stock, and their own has been removed from four to twenty feet, some of the bees may find a hive, but are just as liable to enter the wrong one as the right. Probably they would not go over twenty feet, and very likely not that, unless the new situation was very conspicuous. If a person had but one stock, the loss would probably be less, as every bee finding a hive, would be sure to be right, and none would be killed, as is generally the case, when a few enter a strange hive. Sometimes a stock will allow strange bees to unite with them, but it is seldom, unless a large number enters. When bees are taken beyond their knowledge of country, some two miles or more, the result is somewhat different, but not always without loss, especially if many hives are set too closely together. They leave the hive, of course, without knowing that the situation has been changed; and perhaps get a few feet from it, before strange objects inform them of the fact. When they return, the immediate vicinity is strange, and they often enter their neighbor's domiciles. Experience has satisfied me that stocks should occupy their situation for the summer, as early as possible in the spring, at least before they mark the location; or, if they must be moved after that, let it be *not less than a mile and a half, with plenty of room between the stands.*

SPACE BETWEEN STANDS.

Regarding the distance between hives generally, I would say, let it be as great as convenience will allow. Want of room sometimes makes it necessary to set them closer. Where such necessity exists, if the hives were dissimilar in color, some dark, others light, alternately, it would greatly assist the bees in recognizing their own hive. But it should be borne in mind, that whenever economy of space dictates less than two feet, there are often bees enough lost by entering the wrong hives, to pay the rent of a small addition to a bee-yard. I have several other reasons for recommending plenty of room between hives, which will be mentioned hereafter.

SMALL MATTERS.

The reader who is accustomed to do things on a large scale, will consider so much attention to such a small matter, rather unnecessary, but attention to little things insures success. A grain of wheat is insignificant in itself—it is only in the aggregate that its importance is manifest. The bee is small, the load of honey brought home by it, is still less, and the quantity secreted in the nectary of each flower, still more minute. The patient bee obtains but a tiny drop from each, but by perseverance, procures a load and deposits it in the hive. It is only in the accumulation of such that we find an object worthy of our notice. We are thus taught to look to little things, and the manner in which they are multiplied and preserved. It is much better to save our bees, than to waste them, and wait for others to be raised. "A penny saved is two pence earned." If a stock is lost by a little neglect, a corresponding effort is only necessary to save it. This trifling attention is sometimes neglected through indolence. But I hope for better things generally. I am willing to believe that it is through ignorance of the kind of care necessary, and how, when, and where to bestow it. It seems to be my duty to state the cause of such losses; therefore make it a rule to have stands, bee-houses, etc., ready in spring before the bees leave the hives, and let them remain stationary during the summer.

If we keep bees for ornament, it would be well to build bee-houses, paint the hives, etc.; but as I suspect that the majority of readers will be chiefly interested in the profits, I assure them that the bees will not pay a cent towards extra expenses; they will not do any more labor in a painted house, than if it were thatched with straw. When profit is the only object, economy would dictate that labor be bestowed only where there is a reasonable prospect of remuneration.

CHEAP STAND.

So many kinds of bee-houses and stands have been recommended, all so different from what I prefer, that I perhaps ought to feel some hesitancy in offering one so cheap and simple; but as profit is my object, I shall offer no apology. I have thirty years experience to prove its efficacy, and have no fears in recommending it. I make stands in this way. For a box hive, a board about fifteen inches wide is cut off two feet long; a piece of durable wood two by three inches, is nailed on each end. This raises the board just three inches from the earth, and will project in front of the hive some ten inches, making it admirably convenient for the bees to alight before entering the hive, when the grass and weeds are kept down, which is but little trouble. A separate stand for each hive is better than to have several on a bench together, as there can then be no communication by the bees running to and fro. Also, we are apt to give more room between them; and a board or plank will make more stands when cut in pieces, than if left whole.

I used what is termed a canal bottom board, until I found that it did not pay expenses, hence I rejected it, and succeed just as well. It is generally recommended to prevent robbing, and keep out the moth. It may prevent one hive in fifty from being robbed, but as for keeping out the moth, it is about as good a contrivance in its favor, as need be. I am aware that I differ from most apiarians, in placing the stand so near the earth; less than two or three feet between the bees and the earth, it is said, will not answer any way. I shall not urge the adoption of any rule that I have not tested by my own practice. The objection raised, is the dampness arising from the earth, but I am unable to discover the least bad effect from this cause.

DISADVANTAGE OF STANDING TOO HIGH.

Let us compare advantages and disadvantages a little farther. When the bees approach a hive suspended, or standing on a high bench, two or three feet from the earth, towards evening or on a chilly afternoon—and we have many such in spring—even if there is not much wind, they are very apt to miss the hive and fall to the ground, so benumbed with cold, as to be unable to rise again, and by the next morning are lifeless. On the other hand, if the hive is near the earth, with a board as described, there is no *possibility* of their alighting under it, and if they should fail to reach it, and fall to the ground, they can always creep long after they are too cold to fly, and are thus often able to enter the hive, when they can not use their wings. In this way, enough may be saved in one spring, from a few hives, to make a good swarm. Belonging to different hives, the loss is not perceived, yet as much profit might be realized from them, as if they formed an individual swarm. To such as *will* have them away from the earth, I would say, do adopt some plan to save this portion of your best and most willing servants. Have an alighting board project at least one foot in front of the hive, or a board long enough to reach from the bottom of the hive to the ground, upon which they may crawl up to the hive. Do you want an inducement? Examine carefully the earth about your hives, towards sunset, some fair but windy day in April, when it is chilly towards night, and you will be astonished at the number that perish. Most of them will be loaded with pollen, proving them martyrs to their own industry and your negligence. When I see a bench three feet high, and no wider than the bottom of the hive, and no entrance for the bees, except at the bottom, and as many hives crowded on it as it will hold, I no longer wonder that " bee-keeping is all in luck;" the wonder is how they keep them at all. Yet it proves that, with proper management, bee-keeping is not so precarious, after all.

BEST COVER.

I have taken some pains to ascertain the best protection for hives, from the weather, and have concluded that the cheapest covering is as good as any; any thing that will keep the sun and rain from the top, is sufficient. Covers for each hive, like the bottom board, should be separate, and some larger than the top.

BEE-HOUSE UNPROFITABLE.

I have used bee-houses, but they will not pay, and I have discarded them. They are objectionable on account of preventing a free circulation of air; also, it is difficult to construct them so that the sun may strike the hives both in the morning and afternoon, which is quite essential. If they front the south, the middle of the day is the only time when the sun can reach all the hives at once; this is just when they need it least, and in hot weather the combs are sometimes injured by melting. But when the hives stand far enough apart, on separate stands, it is very easy to arrange them to stand in the sun, morning and afternoon, and be shaded four or five hours in the middle of the day.

We are often quite prodigal in building a splendid bee-house, but we think of economy when we come to put our hives in, and are quite sure to pack them too closely.

SOME WILL HAVE THEM.

Notwithstanding the objections here urged against bee-houses, there will be a few, who, if they keep bees at all, must have them in a house. We will see how far they can be accommodated without seriously diminishing their profits. A bee-house, without any pretension to ornament, generally combines the desired consideration with economy. It is usually straight, and contains but one row of hives. A second and even a third is sometimes added, but when the inconvenience of access to the upper row is considered, together with the disadvantage to the bees, it will confine most economists to the single row. The cheapest form is made by setting posts firmly into the ground, six or eight feet apart, three or four feet from front to rear, and five or six feet high. Cut those on the back enough shorter to give a good pitch to the roof, which may be of boards or shingles. Ten or twelve inches from the ground—not more than that—nail or cut in a shoulder, to support a framework of joists, upon which

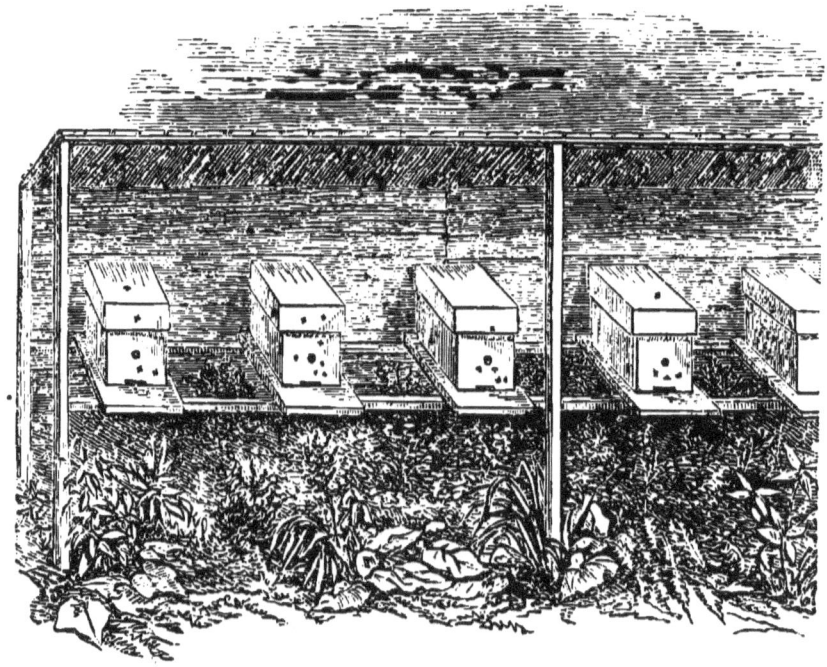

Fig. 14.—BEE HOUSE.

the stands are to be placed. Lay boards the width of the hive or a little more, cross-wise the frame work, and set the hive on the back end.

I consider separate stands, with spaces between, a better arrangement than the plank running lengthwise, as the bees can not run from one hive to the other to gossip. The distance between the hives can be easily regulated; it should be from one to four feet, according to circumstances.

HIVES SHOULD BE OF DIFFERENT COLORS.

The hives should be of different, but not glaring colors. As a bee is guided to the entrance of its home by outside appearances, it is well to alternate the colors when arranging the hives. After the first few days in spring, the workers have but little difficulty. The first day they issue, they seem to settle indiscriminately on the hives to rest, and are often worried and killed. The young bees as they hatch and come out, during the season, either know their own home better, or if they make any such mistakes, are noticed less in the hurry of labor. But the young queen often enters the wrong hive on her return from her excursion, and this uniformly involves the loss of her colony. This is the chief objection to the bee-house.

REPLACING QUEENS.

By using the movable comb hive, and rearing queens artificially, (as the Italian queens are usually raised,) and furnishing laying queens,—thus obviating the necessity of the young queens leaving the hive—this difficulty is avoided. If you think proper, you may allow the bees to swarm, and at the end of a week, look over the combs of the old hive, cut out *all* queen cells, and introduce a laying queen. The after swarms are thus prevented, and the colony is maturing brood nearly two weeks sooner than if they had reared a queen, which is equivalent to a small swarm. Likewise, they will probably be in condition to store surplus, or to part with another good swarm, when they might not have done either, if let alone. This will do much towards balancing the disadvantage of a bee-house. Should you choose to divide—making artificial swarms—the hives should be set far enough apart, *at first*, to allow room for other hives between them. Whenever a colony is sufficiently strong to divide, one half the combs may be put into the new hive, as in making artificial swarms. In a week, cut out cells, and introduce the laying queen. It will be important to keep a few queens on hand, in case any should unexpectedly swarm; or rather you should expect some to do so, a little before you are ready to divide. The advantages of having laying queens always ready, amply repays all trouble in rearing. I would suggest that the movable comb hives only, are used for this kind of bee-house, as it is evident that the box hive can

Fig. 15.—BEE HOUSE.

not be so easily managed. When such a hive swarms, we often can not remove all the queen cells; the bees would be obliged to raise their own queens, and would be likely to swarm several times. Also, the queen left in the old hive, and those with the after swarms, would be likely to be lost.

THE APIARY

SEVERAL BEE-HOUSES.

If one uses nothing but the box hive, and must have bee-houses, it would be better to make several small ones,

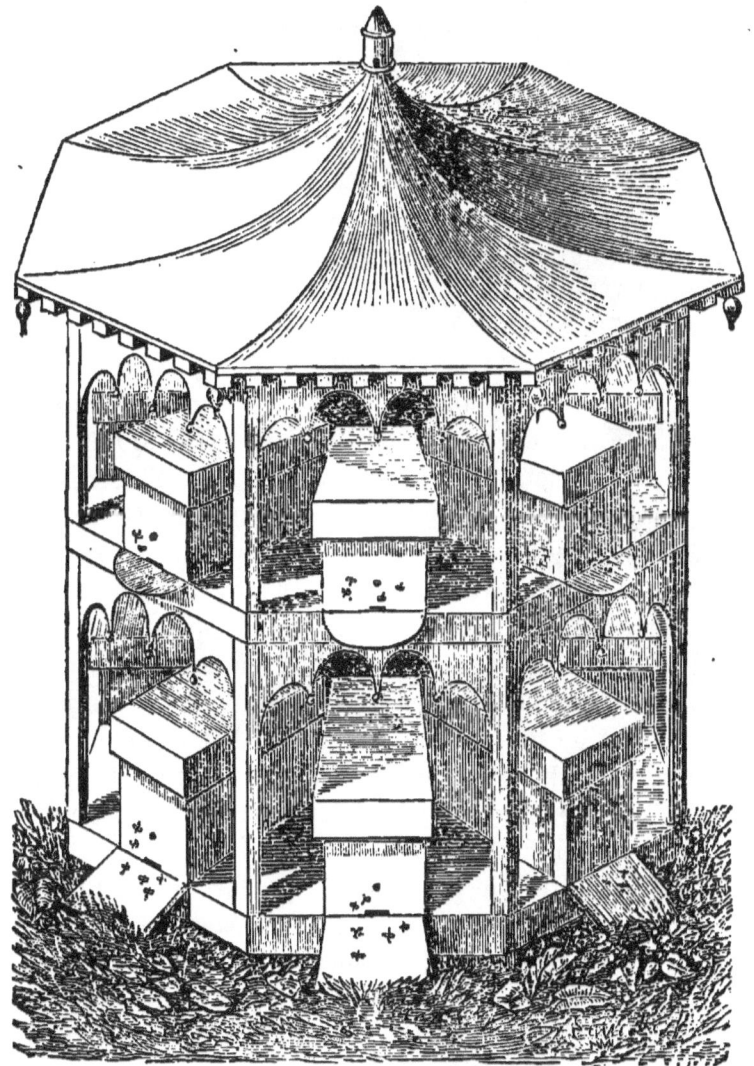

Fig. 16.—BEE HOUSE.

holding but three hives, fronting in three different directions, as in fig. 15. Let it be closely boarded on the north side, or so as to break off the prevailing wind, with a roof over the whole.

It will be seen that the three hives set very compactly, yet the entrances are some distance apart, and so dissimilar that the young queens would seldom make a mistake in entering. In a place where there is little or no wind, a hive may be added on the fourth side. Such bee-houses, tastefully built, would be quite ornamental. A hexagonal shape might be a little more graceful, yet it would increase the liability of losing the queens. For a more elaborate style, an octagon would be suitable, to which a second story might be added, giving room for sixteen hives, as seen in fig. 16, on the preceding page.

In grounds where the bees would annoy the family, or visitors, they may be surrounded by a hedge of shrubbery or vines. Enclose a plot of the desired area, of any shape the fancy may dictate. Should a circle be chosen, I can easily imagine that a small, slender tree in the centre, would greatly assist the bees in finding the hives. The hedge may be evergreens, vines, or any small shrub of suitable growth. Grape vines are very appropriate, being of quick growth, and combining ornament with profit. Openings should be left at proper distances for the hives, which should be set in such a manner that the front is on the inside, and the body of the hive on the outside of the hedge. The hedge should be dense, but as narrow as possible. One arrangement is shown in fig. 17.

Operations with the hives can be performed on the outside, thus avoiding the attacks of the bees at work, which are more liable to be troublesome than those which leave the hive when it is opened. This hedge can be trained in an arch over the top of the hives, but should not be allowed to attain any great height, compelling the bees to rise over it. The fronts of the hives should be of different colors, as before mentioned, but the color of the other parts may be uniform, if desired. Proper openings should be left for the operator to pass within, when necessary.

Fig. 17.—HIVES ARRANGED IN A HEDGE.

Those who consider the appearance of the hive unsightly, may set an outer hedge, a few feet from the first, which will effectually screen them from observation.

CHAPTER VI.
ROBBING.
NOT UNDERSTOOD.

Robbing is often a source of loss to the careless apiarian. It is frequent in spring, and at any time in warm weather, when there is a scarcity of honey. It is very annoying, and is sometimes a source of contention among neighbors, when perhaps neither is to blame, farther than for ignorance. The person keeping the most bees, must expect to be held accountable for all the losses in the neighborhood, whether they occur from mismanagement, or want of management, and if he escapes without being charged with those lost by hundreds of other causes, he ought to be thankful. It is often thought if a person has but one stock, and another has ten, that the ten will combine to plunder the one. This conclusion is not warranted by facts. I can discover no collusion between different families of the same apiary. It is true that when one colony finds another weak and defenceless, possessing treasures, they have no conscientious scruples about carrying off the last particle, notwithstanding they revel in abundance at home; and it is most frequently the case that the strongest colonies are most given to this despicable habit. The hurry and bustle attending the plunder, seldom escapes the notice of other hives, and when one hive in the yard has been robbed, perhaps two-thirds or all of the others have participated in the offense.

It is common to hear remarks like this, "I had a *first rate* hive of bees," (when, in fact, he had not looked at them, particularly, for a month, and knew nothing of their real condition)—" and Mr. A's bees began to rob them." I tried every thing to stop it; moved them several times to prevent their finding the hive, but it did no good; the first I knew they were all gone—bees, honey and all! The bees all joined the robbers." Now, the fact is, that not one good colony in fifty will ever be robbed, if let alone; that is, if the entrance is properly protected. Moving the hive was enough to ruin it; bees were lost at every change, until nothing was left but honey to tempt the robbers; whereas, if left on its stand, it might have escaped.

The injury done by robbers is sometimes like that done

by worms, and usually following some preceding weakness of the colony. Not one *strong* colony in a hundred will be attacked and plundered at the first onset.

DIFFICULTY IN DECIDING.

Probably but few bee-keepers are able to decide at once *when bees are robbing.* It requires the closest scrutiny to decide. There is nothing about the apiary more difficult to determine; nothing in which one is more likely to be deceived. It is generally supposed, when a number are fighting outside, that it is conclusive that they are also robbing, which is seldom the case. On the contrary, a show of resistance indicates a strong colony, and that they are disposed to defend their treasures. A very weak colony of Italians will often make a spirited resistance. I have no fears for a stock that has courage to repel an attack. The greatest danger is with those weak colonies incapable of opposition. In seasons of scarcity, all *good* stocks maintain sentinels about the entrance, whose duty it appears to be to examine every bee that attempts to enter. If it is a member of the community, it is allowed to pass; if not, it is arrested on the spot. It would seem that a password was requisite for admittance, for no sooner does a stranger endeavor to enter, than it is known. The absence of proper credentials is evidence enough to convict it. Each bee is a qualified jurist, judge, and executioner. There is no delay, no waiting for witnesses for the defence. The more a bee attempts to escape, unless it is by chance successful, the more certain is the execution of the sentence. How strange bees are known to be such, is yet undetermined, probably by the scent.

WEAK COLONIES IN DANGER.

It is the duty of every bee-keeper who expects to succeed, to know which his weak stocks are. An examination can be made on some cool morning, by turning the hive bottom up, and allowing the sun to shine among the combs. The number of inhabitants is thus easily seen. When weak, close the entrance till there is just room for one bee to pass at once.

WHEN TO LOOK OUT FOR ROBBERS.

A little after noon, on the first pleasant day, at any time before honey is obtained plentifully, look out for robbers. To get to robbing, bees must be first tempted, and rendered furious. A dish of refuse honey left near them is sometimes sufficient to set them at work; also an insufficient supply, when fed. After they have once commenced, it takes an astonishing quantity to satiate their appetite. They seem to be perfectly intoxicated, and reckless of danger, venturing into certain destruction. I have known a few instances where good colonies were so reduced by fighting while robbing, that they in turn fell a prey to similar marauders.

I have for several years kept hundreds of stocks away from home, where I could seldom see them. Yet I seldom lose a stock by robbing. I simply keep the entrance closed, leaving, during spring, a passage for the bees at work. It is true, I have lost a few, when the other bees took the honey, but they would have been lost in any case. A great many apiarians raise their hives an inch from the board, early in spring. They seem to disregard the opportunity it gives robbers to enter on every side. It is like setting the door of your dwelling open, to tempt the thief, and then complaining of the consequences.

Let it be understood, then, that all good stocks, under ordinary circumstances, will take care of themselves. Nature has provided means of defence, with instinct to direct its use. Non-resistance may do for highly cultivated intellect in man, but not here. There is a prevalent opinion that robbers often go to a neighboring hive, kill off the bees first, and then take possession of the spoils. I have never yet discovered one fact to corroborate this, although I have watched very closely. Whenever bees have lost all their stores, at a period when nothing was to be obtained from flowers, it is evident that they must soon starve, and disappear in a few days. This would naturally give rise to the supposition that they had been killed by the robbers.

FIRST INDICATION.

I will now describe the appearance of a weak hive that is being robbed, and show, that without timely interference, the result will be a total loss to the colony. Each robber, when leaving the hive, instead of flying in a direct line to its home, will turn its head towards the hive to mark the spot, that it may return for another load, in the same manner that they do when leaving their own hive for the first time in the spring. The first time the young bees leave home, they mark their location by the same process. A few of these begin to hatch very early, in all good stocks, often before the weather is warm enough for any to leave the hive. These young bees will fly out very thickly about the middle of each fair day, or a little later, called "playing," by some writers. This unusual activity strongly resembles the bustle of robbers, and it is difficult to detect the difference. Their motions are alike, but there is a little difference in color, the young bees being a shade lighter; and the abdomen of the robbers, when filled with honey, is a little larger. But while you are learning these nice distinctions, your bees may be ruined. I will therefore give additional means of ascertaining. Bees, when they have been stealing a sack of honey from a neighboring hive, will generally run several inches from the entrance before flying ; kill some of these ; if filled with honey, they are robbers; for it is very suspicious to be filled with honey when leaving the hive; or sprinkle some flour on them as they come out, and let some one watch by the other hives to see if they enter. The following is less trouble, but it will be longer before they are checked, if robbing. Visit them again in the course of half an hour or more, after the young bees have returned, and if the bustle continues or increases, it is time to interfere. When the entrance has been contracted, as directed, close it entirely, till near sunset. If it has been left open, it should now be closed, giving room for only one bee at a time. This will allow all that belong to the hive to get in, and others to get out, and will materially retard the progress of the robbers. Unless it should be cool, they will continue their operations till evening. Very often some are unable to get home in the dark, and are lost. This, by the way, is another good test of robbing. Visit the hives every warm evening. They commence depredations on the warmest days, seldom at any other time. If any are at work when honest laborers should be at home, they should be regarded with suspicion.

REMEDIES.

A great many remedies for this evil have been recommended, which are as bad as the evil itself, and often the cause of it. The most fatal is to move the hive a few rods; another, to entirely close it, which may smother the bees; or to break out some comb in the hives of the robbers, and set the honey to running, thereby giving them work at home. I would recommend removing the weak hive, on the morning after the attack, to the cellar, or some dark, cool place, until two or three warm days have passed, and the search has been abandoned. The robbers will probably attack the stock on the next stand. Contract the entrance of this according to the number of bees that are to pass. If strong, no danger need be apprehended; they may fight and even kill some; perhaps a little chastisement is necessary to bring them to a sense of their duty. When a hive has been removed, if the one on the next stand is weak, it is better to take that in also, to be returned as soon as the robbers will allow it. If a second attack is made, put them in again, or if practicable, remove them a mile or two, out of their knowledge of country; they would then lose no time from labor. Where but few hives are kept, and not more than one or two are engaged, sprinkle a little flour on the bees as they leave, to ascertain which hive they are from ; then reverse their position, putting the robbed in the place of the robbing, and the reverse. The weak hive will generally become the strongest, and put a stop to their operations. But this method is impracticable in a large apiary, because several stocks are usually engaged very soon after one begins, and a dozen may be robbing one. Another method is, when you are *sure* a stock is being robbed, to close the hive at a time when there as many plunderers inside as possible, (wire-cloth, or something that will admit air, and confine the bees, is necessary,) and carry in as before directed, for two or three days, when they may be set out again. The strange bees thus enclosed will join the weak family, and will be as eager to defend what is now *their* treasure, as they were before to carry it off. This principle of forgetting home, and uniting with others, after a lapse of a few days, can be wisely acted upon in this case. It succeeds about four times in five, when a proper number is enclosed. Weak families are very easily strengthened in this way, and the bees, being taken from a number of hives, are scarcely missed elsewhere. The difficulty is, to secure about as many as belong to the weak hive; if too few are enclosed, they are apt to be destroyed. As I remarked in the beginning of the chapter, bees will plunder and fight at any time through the summer when honey can not be collected; but *spring* is the only time in which such desperate and persevering efforts are made to obtain it. At no other time can the apiarian be excused for having his hives plundered, or allowing them to be liable to such invasions. Families reduced in winter and spring, will, if protected through this season, generally make good stocks. Prevention is better than cure; evil propensities should be checked in the bud. It would probably be the least trouble, when practicable, to remove the weak hive to some neighbor, a mile or two away, where there are no strong ones to molest it; and return it after the honey season arrives.

The apiarian who allows his hives to be plundered in the fall, is not fit to have charge of them; the efforts of the robbers are seldom as vigorous as in spring, (unless there is a general scarcity,) the weak hives are usually better supplied with bees, and consequently a less number is exposed. When there are some very weak families, they should be disposed of *as soon as the flowers fail*. Particulars given in Fall Management.

EQUALIZATION.

I have sometimes equalized the strength of my hives, early in spring, by the following method. Bees, when wintered together in a room, will seldom quarrel when first put out. When one hive has an over supply of bees, and another a very few, the next day after being set out, I change the weak one to the stand of the strong one, (as before mentioned), and all bees that have marked the location, will return to that place. This often fails for the reason that too many bees leave the strong hive, making that the weak one, and nothing is gained. If it could be done when they had been out of winter quarters just long enough for the proper number to have marked the location, success would be quite certain. But before an exchange of this kind is made, it would be well, if possible, to ascertain the cause of a stock being weak; if it arises from the loss of a queen, we only make the matter worse by this operation.

BATTLES.

I will describe some of their battles, or what are called battles, as it is seldom that a regular battle occurs, in which both parties make a deadly effort to destroy the other. Two queens will meet thus, and occasionally two workers. Bees fight to repel invaders, but I have little faith that they make war on a neighboring colony for the mere sake of fighting. When bees first fly out in spring, some will settle on a neighboring hive, if they are close together, but as soon as one alights, it is surrounded, the whole front of the hive being sometimes covered in this way. A half dozen will attack one stranger, two or three biting its legs, one pulling it by the wing, another perched on its back making a feint of stinging, while another is ready to take what honey it has, when it has been worried sufficiently to give it up. It is sometimes let go, after it has yielded all its honey, but is often dispatched by a sting, which is almost instantly fatal. A bee is killed by a sting sooner than by any other means, except crushing. When strange bees enter a hive, which sometimes part of a returning swarm will do, I have known a few thousands to be killed in five minutes. The joints are the only vulnerable parts of a bee. During the fight, if the object be to repel pillagers, a few bees may be seen buzzing around in search of an unguarded place to enter the hive. If such is found, it alights and enters in a moment. At other times, it meets a sentinel on duty, and is on the wing again, in an instant. It is occasionally more unfortunate, and is seized by the guard, when it must either break away, or suffer the penalty of insect justice, which is generally "to the utmost extent of the law."

CHAPTER VII.

FEEDING.

FEEDING A LAST RESORT.

Feeding bees is sometimes quite necessary. But in ordinary seasons and circumstances it is very doubtful policy for the apiarian to attempt to winter many stocks so poorly supplied with honey, that they will need feeding before spring. Nothing is more common than for inexperienced persons to undertake to winter every hive containing bees, and the more ignorant they are of the business, the more poor hives they will endeavor to keep. There are circumstances under which it may be proper to feed colonies in the fall. In the chapter on Fall Management, I will give directions for disposing of such as should not be fed. It pays better to feed in spring than at any other season, and there are more that need it then. Some families having had light stores at the beginning of winter have consumed about all. Some stocks, when brought from their winter quarters, mix badly with others, and occasionally most of the bees leave their own hive, and join other stocks. Those left may not be able to defend their stores, and will be robbed.

I have known a few instances where there was every requisite for a good stock, and yet they were so imbecile that they would not defend themselves, and allowed every particle to be taken from them. Although there is a strong temptation to let such starve, as a punishment for want of energy, it usually pays to feed them. Bees may also be fed at this season to promote early swarming and storing of surplus honey.

CARE.

In feeding, the utmost care is requisite, and but few know how to manage it properly. Honey fed to bees, is almost certain to excite quarreling. Strong colonies sometimes scent the honey given to weak ones, and carry it off as fast as supplied. It is possible that feeding a stock of bees in spring, may cause them to starve; whereas, if let alone, they would survive. Notwithstanding this seems contradictory, it may be made to appear reasonable. Whenever the supply of honey is deficient, probably not more than one egg in twenty will be matured, their means not allowing the brood to be fed. In very small colonies the queen usually confines herself to a small area of comb, often depositing several eggs in one cell, but if the supply of honey is increased, she will extend her labors over a greater space. Suppose we give such a stock two or three pounds of honey, encouraging them to feed a large brood, and the supply fails before they are half grown. What are they to do? Destroy the brood and lose all they have fed, or draw on their old stores, and trust to chance for themselves? The latter alternative will probably be adopted, and then without timely intervention of favorable weather, the bees will starve. The same effect is sometimes produced by the changes of the weather. A week or two may be very fine and bring out the flowers in abundance, and a sudden change, perhaps frost, may cut them all off. This makes it necessary to exercise considerable vigilance, as these spells of cold weather make it unsafe to neglect them, till white clover appears, (10th or 15th of June in this section) but if the spring is favorable, there is but little danger, unless they are robbed as fed. If the necessary care be taken about moth-worms, the light hives can be distinguished. This is another advantage of the simple hive; by merely raising one edge to destroy worms, we learn something about the quantity of honey on hand. To be very exact, the hive should be weighed when ready for the bees, and the weight marked on it; by weighing at any time after, we can ascertain pretty nearly the amount of honey. Some allowance must be made for the age of the combs, quantity of brood, etc. It is wrong to begin to feed without being prepared to continue, as the supply must be kept up until honey is abundant.

DESTITUTE COLONIES SOMETIMES DESERT.

When one has the means to continue feeding, and time requisite to make it secure, perhaps it would not be advisable to wait till the last extremity before feeding, as a small family will sometimes entirely desert the hive, when destitute, especially if they have but little brood. In these cases, they issue precisely as a swarm; after flying a long time, they either return or unite with some other stock, but seldom cluster. If they return, they need attention immediately, and we may be certain there is something wrong, let the desertion take place when it may. In spring, the cause may be destitution, or mouldy combs; at other times, the presence of worms, diseased brood, etc. But whatever the cause, ascertain it, and apply the remedy.

WHEN THEY MUST BE FED.

If it is wished to wait as long as possible before feeding, a test will be necessary to decide how long it will do to delay it. *Strict attention must be given; they will need examination every morning.* If a light tap on the hive is answered by a lively buzzing, they are not suffering yet; but if no answer is made to this inquiry, it indicates weakness. Extreme destitution takes away all disposition to repel an attack. Sometimes a part of the bees will be too weak to remain among the combs, and will be seen lying on the bottom, while a few will be outside. If the weather is cool, they will be apparently lifeless; yet they can be revived, and now *must* be fed. Those among the combs may be able to move, though feebly. When this is the condition of things, invert the hive, gather up all the scattered bees, and put them in. Get some honey, if candied, heat it till it dissolves; comb honey is not so good unless broken up; pour a quantity among the combs, directly on the bees; cover the bottom of the hive with a cloth, securing it firmly, and bring to the fire to warm. If no honey is at hand, sugar may be used instead; add a little water, boil until near the consistency of honey, and skim it; when cool enough, use the same as honey. In two or three hours they will be revived, and may be returned to the stand, providing the honey given is all taken up; on no account let any run out around the bottom. The necessity of a daily visit to the hives is apparent from the fact that, if left for only a day or two in the situation just described, it will be too late to revive them.

At night, if you have a box cover, such as recommended, you may open the holes in the top of the hive, fill a dish with honey or syrup, and set it on the top; put in some shavings, cut straw, or a float made of very light wood, very thin, and full of holes, or narrow channels made with a saw, to keep the bees from drowning. When you begin to feed, scatter a few drops on the top of the hive, down into it, and on the side of the dish to teach them the way. When the weather is warm enough for them to take it during the night, it is best to feed at evening—from four to eight ounces daily is sufficient. If the family is very small, what honey is left in the morning, may be taken by robbers. It is then best to take it out, or carry the hive into a dark room, sufficiently warm, and feed them enough

FEEDING.

to last several days, and then return them to the stand—keeping a good look out that they are not plundered, and again starving, until the flowers produce sufficient honey.

MANNER OF FEEDING.

The following is a more systematic mode of feeding. Get a tinman to make a dish, ten or twelve inches square, with vertical sides two inches high. For a box hive cut a board two feet long, and fifteen inches wide; two or three inches from one end, cut out a place exactly the size of the dish, so that it will set in just even with the upper side of the board. Make a good fit, that no bees may get in around it. Nail cleats on the under side one or two inches thick. To keep the bees from drowning when the dish is filled with honey, and to prevent them from making

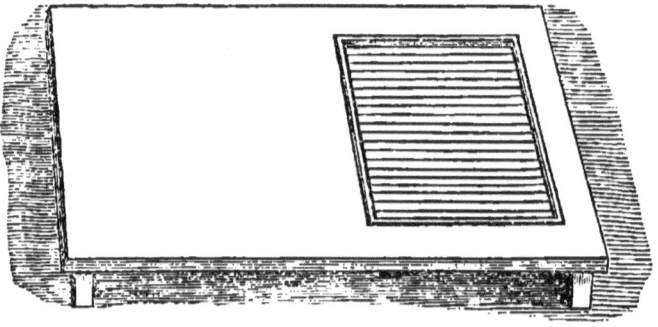

Fig. 18.—FEEDER.

combs down into it, set in some thin strips edgewise, half an inch apart, and reaching nearly to the bottom. To hold these strips in place, put a piece of half inch board, two inches wide, across each end. With a thick or coarse saw, cut channels half an inch apart in one side of these pieces, one-fourth inch deep, and crowd the thin strips into them even with the top of the dish.

The strips may be split out of shingles, or sawed for the purpose. Set the hive over this, leaving one end of the dish two inches outside the back of the hive, for convenience in filling. Lay over this a strip of board to keep out the bees. If the weather is warm when feeding, the bees will soon get cross, unless smoke is instantly blown among them on raising the cover. Bees will take honey more readily when directly under them, than when overhead, or on the side. Yet for most purposes the latter places will answer. To feed at the back side, make a shelf for the dish described, and a frame an inch deep, just the size of the dish. Make two or three holes in one side, and corresponding ones in the side of the hive. Lay this frame on the dish, with the holes next the hive, and put over it a pane of glass. The bees can enter from the hive, and no outsiders can interfere.

Whatever plan of feeding is adopted, all openings large enough to admit a bee, except the regular entrance, should be closed. The board and feeder can be taken away, when feeding is over, and put aside until needed again. If left under the hive through the summer, it affords rather too convenient a place for the worms to spin their cocoons.

OBJECT IN FEEDING.

If the object in feeding is to induce early swarming, of course the best stocks are to be chosen for this purpose; but care is necessary not to give them too much, and thus have the combs filled with honey, that should be occupied with brood, thereby defeating your object. One pound per day is enough, perhaps too much. The quantity obtained from flowers is a partial guide; when plenty, feed less, when scarce, more. Begin as soon as they will take it up in spring, and continue in accordance with the weather, until white clover blossoms, or swarms issue. Another object in feeding bees at this period, is to have the store combs all filled with inferior honey, so that when clover appears, there is no room for it, except in the boxes, which being now put on, are rapidly filled. Inferior honey may be used for this purpose; Southern or West India is good, and of moderate cost. Inferior sugar, mixed with honey, will do, but they do not relish it so well when fed alone. I have usually taken about equal quantities of each, adding a pint of water to ten pounds of the mixture, boiling and skimming it. The idea has been advanced that candied honey is injurious to bees—even fatal. I never could discover any unfavorable result, further than that it was a perfect waste, when fed in this state. When boiled, and a little water added, it appears to be as good as anything. Nearly every stock will have more or less of it on hand at this season, but as warm weather approaches, and the bees increase, it seems to become liquified from heat alone. The bees, when compelled to use honey thus candied, waste a large portion; a part is liquid and the rest is grained like sugar, which may be seen on the bottom board as the bees throw it out.

Another and less commendable object in feeding bees, is to give inferior honey, mixed with sugar, and flavored to the taste, to the bees, and let them store it in boxes for market. I have no faith that honey undergoes any chemical change in the stomach of the bees, while they are going from the feeding dish to deposit it in the cell, and can not recommend this as an honest course. Neither do I think it would be very profitable to feed for this purpose, under any circumstances. I have sometimes had boxes nearly full, and almost ready for market, at the end of the honey season; when it would seem that feeding a little would complete them, provided the hive were full. I have fed them a few pounds of good honey at such times, but I always found that of several pounds fed, but very little would be stored in the boxes.

PROMISCUOUS FEEDING UNPROFITABLE.

I have seen it recommended and practiced by some apiarians, to feed bees all at once, in the open air, in a large trough; but whoever realizes much profit from this method, will be peculiarly fortunate, as every stock in the neighborhood will soon scent it, and carry off a good share. Also, nearly every stock at home will be in contention, and great numbers be destroyed. The moment the supply is exhausted, their attention is directed to other stocks. Another objection to this wholesale feeding is, that some stocks do not need it at all, while others do, and the former, being stronger, are quite likely to get the most.

CHAPTER VIII.

DESTRUCTION OF THE MOTH-WORM.

I shall not give a full history of the moth in this place, as spring is not the time in which it is most destructive. But as this is a duty belonging to spring, a partial history seems necessary.

As soon as the bees begin their labors the worms are generally ready to commence theirs.

FOUND IN THE BEST STOCKS.

You will probably find some in your best stocks, but it need not alarm you. Even weak colonies are seldom destroyed at this season, although all may be more or less injured. They work mostly among the sealed brood. The heads of the young bees, after assuming the chrysalis form, are about one-twelfth of an inch from the sealing of the cells, leaving a suitable space between their heads and the sealing, for the perambulations of the worms.

FEAR OF THE BEES.

As a protection against the bees, they spin silken galleries, completely surrounding them, and never exposing any part but the head, which is coated with mail. They are thus safe until they have attained their growth, when

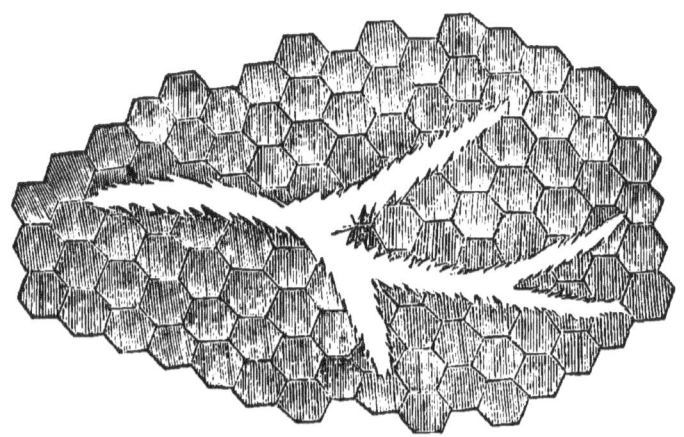

Fig. 19.—WORM GALLERY IN THE COMB.

it is necessary to leave their feeding grounds to find a place for their cocoons. Without this silken covering, they are easily annoyed by the legal occupants of the hive, and will creep into every available corner to avoid their attacks. During the cool nights of spring, they become chilled, get on the bottom board, and being unable to move further, are easily found and killed. If you have the movable comb hive, you can take out the frames, and trace this silken gallery from beginning to end. Touch it in different places with a sharp pointed knife, till you see by a stir inside where the worm is, then with the point of the knife, and thumb, it can be picked out at once. To destroy such as have left the combs, get a piece of narrow hoop-iron, (steel would be better) three-

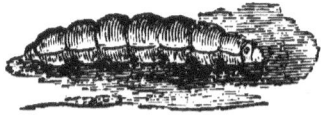

Fig. 20.—MOTH-WORM.

fourths of an inch wide, and five inches long; taper one side three inches from the end, to a point, then grind each edge sharp, make three or four holes through the wide

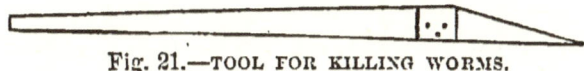

Fig. 21.—TOOL FOR KILLING WORMS.

end, to admit small nails through it in the handle, which should be about two feet long, and half an inch square. Armed with this you can proceed.

HOW DESTROYED.

Raise the hive on one edge, and with the point of your sword you may pick a worm out of the closest corner, and easily scrape all from under the hive. *Be sure and dispatch every one;* not that the "little victim" will personally do much more mischief, but it is to be apprehended through its descendants. Very likely half of all you find will have finished their course of destruction among the combs, and voluntarily left them for a place to spin their cocoons. They are worried, if bees are numerous, until satisfied that there is no safe place among them to make a shroud and remain helpless for two or three weeks. Accordingly, when they get their growth, they leave and get on the bottom board. They will be chilled and helpless in the morning, but active in the middle of the day. If they are merely thrown on the earth, a place will be selected there for transformation, if no better is found, and a moth perfected ten feet from the hive, is just as capable of depositing five hundred eggs there, as if she had never left it.

Several generations are matured in the course of one summer; consequently, one destroyed at this season, may prevent the existence of thousands before the summer is over.

The moth-worm is one of the many subjects connected

with bees, concerning which, there is a great deal of theoretical reasoning and imposition. I wish the reader to judge for himself, lay aside whims and prejudices, and look at the subject candidly; and if no testimony comes up to confirm any position I assume, I shall not complain if my assertions fare no better than some others. Only defer judgment until you *know* for yourself.

Bees have always received my special regard and attention, and my enthusiasm may blind my judgment. I may be prejudiced, but not wilfully wrong. I have found so many theories utterly false, when carried into practice, that I depend on none, however plausible, without facts to support them.

To return to our subject. It is supposed by many, when these worms are found on the board, that they get there by accident, having dropped from the combs above. They do not seem to understand that the worm generally travels on safe principles, that he attaches a thread to whatever he passes over. To be satisfied on this point, I have many times carefully detached his foot-hold, when on the side of the hive, or other place, where he would fall a few inches, and always found him with a thread fast at the place he left, to enable him to regain his former position if he chose. Is it not probable, then, that whenever he leaves the combs for the bottom board, he can readily ascend again? No doubt he often does, to be driven down again by the bees. Now, what I wish to show by all this preamble, is simply this; that all our trouble and anxiety to prevent the worms from again ascending to the combs, by wire hooks, wire pins, screws, nails, turned pins, clam shells, blocks of wood, etc., is perfect nonsense, when half of them would do the bees no harm if they did return, and might as well go there as any where else. And, these useless contrivances are very often positively injurious to the bees.

Suppose, if you please, that the worm has no thread attached above, and the hive is far enough above to prevent his reaching it. Of course he can't get up; but how are the bees to do any better? The worm can reach as high as they can. You think the bee can fly up; so it will, sometimes, but will try a dozen times to get up some other way, and when it does fly, a smooth board is a very bad place from which to start. Did you ever watch by a hive thus raised, towards night, in April or May, when it was a little cool, and see the industrious little insects arriving with a load as heavy as they could possibly carry, chilly, and nearly out of breath, scarcely able to reach home, and there witness their vain attempts to get among their fellows above them? If you never observed this, I wish you would do so, and when you find them giving up in despair, and perishing after many fruitless attempts, I think if you possess sympathy, benevolence, or even selfishness, you will be induced to do as I did—discard at once wire hooks, and all other contrivances under the hive, in the spring, and give the bees, when they do get home with a load, what they richly deserve,—protection.

But if you set the hives close to the bottom board, you will say "the worms will get between the bottom of the hive and the board." Well, what then? I expect if you intend to succeed, that you will get them out, and destroy them. I am as willing to find a worm under the edge of the hive, and dispatch it, as to have it creep into some place out of sight, and change to a moth. I once trimmed off the bottom of my hives to a thin edge, so they could not have this place for their cocoons, but I now prefer to have them square. No investment brings profitable returns without proper attention. If you plant a field with corn, you do not expect that the whole work is finished with the planting. Neither should you expect when you set up a stock of bees, that a full yield will be realized without further care. If you are remunerated for keeping the weeds from your corn, be assured that it is equally profitable to weed out your bees.

Now, do not be deceived, and through indolence be induced to get hives with descending bottom boards, to throw out the worms as they fall, and hope by that means to have no further trouble. We will suppose such inclined bottom board capable of throwing every worm that touches it "heels over head" to the ground; what have we gained? His neck is not broken, nor any other *bone* of his body! As if nothing extraordinary had happened, he quietly gathers himself up and looks about for snug quarters; he cares not a fig for the hive now; he gormandized on the combs until satisfied before he left them, and is glad to get away from the bees at any price. A place large enough for a cocoon is easily found, and when he again becomes desirous of visiting the hives, it is not to satisfy his own wants, but to accommodate his progeny.

MOTH PROOF HIVES NOT MADE.

A hive that is proof against the moth is yet to be constructed. We frequently hear of it, from patent-venders, but when tested by practical bee-keepers, the worms are generally found in the vicinity of the bees. When your hives become so full of bees that they cover the board in a cool morning, you will seldom find the worms, except under the edge of the hive. You may now raise it, and catch the worms by laying under the bees a narrow shingle, a stick of elder split in two, lengthwise, with the pith scraped out, or any thing else that will afford them protection from the bees, and where they may spin their cocoons. These should be removed every few days, the worms destroyed, and the traps put back. Do not neglect it till they change to the moth, and there is nothing to remove but empty cocoons.

BOX FOR WREN.

If you would take the trouble to put up a cage or two for the wren to nest in, he would be a valuable assistant in this department of your labor. He would be on the lookout when you were away, and many worms, while looking up quiet lodgings, would be relieved from all further trouble by being deposited in his crop. The cage need not be more than four inches square, and should be fastened as near as possible to the bees, to a post, tree, or side of a building, a few feet from the ground. The skull of some animal, (horse or ox) is very convenient for them, the cavity for the brains being used for the nest. A person once told me the wren would not build in one that he had put up. On examination, the stake to support it was found driven into the only entrance. I mention this to show how little some people understand what they are doing. It is sometimes as well to know *why* a thing is to be done, as to know it *must* be done. If this prolixity is unnecessary for one, it may benefit another. You must remember that some bee-keepers are not over supplied with ingenuity, and must receive very explicit directions.

CHAPTER IX.

PUTTING ON AND TAKING OFF BOXES.

MUST NOT BE PUT ON TOO EARLY.

Putting on boxes may be considered a duty intermediate between spring and summer management. I can not recommend putting them on, in ordinary circumstances, as early as the last of April or first of May. Before the hive is full of bees, it is generally useless, quite likely a disadvantage, by allowing a portion of animal heat to escape, which is needed in the hive to mature the brood. Also, moisture may accumulate in them, causing mold to form on the inside. Experience and judgment are necessary to know about what time boxes are needed. That they are necessary, need not be argued at the present day. Bee-keepers have generally abandoned the barbarous practice of killing the bees to obtain the honey. Most of them have learned that a good swarm will store sufficient honey for winter, besides several dollars' worth of surplus. Here is where the patent-vender has taken advantage of our ignorance, by pretending that no hive but his, *ever obtained such quantities of honey, and of such pure quality.* It is probable that a great many bee-keepers are unable to tell precisely when the hive is full of honey; it may be full of bees, and not of honey, and they are thus uncertain when to put on boxes. The best rule that I can give, is to put them on when the bees begin to be crowded out. When they are obtaining honey, a day or two before this, would be just the right time. It should be remembered that they do not always get honey when they begin to cluster outside. This guide will do in place of a better one, which only close observation and experience can give.

You may discover whether they are obtaining honey by attentively watching the cells next the glass in a glass hive. If honey is being deposited there in abundance, it is quite evident that the flowers are yielding it, and other stocks are obtaining it also. Now is the time, if the hives are full, to put on the boxes. Too much room might retard the swarming a few days, but if the bees are crowded outside, it indicates want of room, and the boxes can make no difference. It is better to have one box well filled, than two part full, as might be the case if the bees were not numerous. The object of putting on boxes before swarming, is to employ a portion of the bees, that otherwise would remain idly clustering outside for two or three weeks, as they often do while preparing the young queens for swarming. But when all the bees can be profitably engaged in the body of the hive, more room is unnecessary.

MAKING HOLES AFTER THE HIVE IS FULL

If a hive has no holes in the top, it need not prevent your getting a few pounds of pure surplus honey.

If the holes are about two inches apart, and the row is at right angles with the combs, each one will be so made that a bee arriving at the top of a hive, between any two combs, will be able to find a passage into the box without long search, which I can imagine he would have when only one hole is made, or when they are parallel with the combs. If a hive contained eight or ten sheets of comb, and but one passage to the box, a bee might go up between any two, many times, before it found the opening.

It has been urged that every bee soon learns all the passages about the hive, and consequently will know the direct road to the box. This may be true, but when we recollect that all within the hive is perfect darkness—that the sense of feeling must guide the bee in all its travels, and that perhaps a thousand or two young workers are added every day, and these have to learn by the same means, we would, if we studied our own interest, give them all possible facilities for entering the boxes. What way so easy for them as to find a passage, when they get to the top, between each two combs? That bees do not know all roads about a hive, can be partially proved by opening the door of a glass hive. Most of the bees about leaving, instead of going to the bottom to make their exit, seem to know nothing of the way, and will vainly try to get out through the glass. I am so well convinced of this, that I take some pains to accommodate them with frequent passages.

To assist them as much as possible, when swarms are put in new hives, make the holes and use guide-combs, as directed for boxes, which should cross the row of holes at right angles.

To make holes in the top of full hives, mark out the top as directed for making hives and boxes. A centre bit, or an auger bit, with a lip or barb, is best, as that cuts down a little faster than the chip is taken out, leaving it smooth. When nearly through, you can cut the remainder of the chip loose, with a pointed knife, and take it out. If it is between the combs it is well; if directly over one, better; with the knife take out a piece of comb as large as a walnut. The bees will then have a passage through from either side of the comb.

After you have opened one hole, the bees will very likely want to know what is going on over head, and send out a force "to make a reconnoissance." To prevent their interference, use smoke, and send them down out of your way, till the hole is finished; then lay over it a small stone or block of wood, and make the others in the same way. When all are done, blow in some smoke as you uncover them, and put on the boxes. This process is not half so formidable as it appears from the description.

BOXES MAY BE TOO EASY OF ACCESS.

Dr. Bevan and some others have made a cross-bar hive, by laying strips of half-inch board, a little over an inch wide, and half an inch apart, across the top, instead of nailing on in the usual way. It is plain that in such a hive, a bee can pass into the box without difficulty, whenever it arrives at the top. I will repeat my objection to allowing too much room to passages into the boxes, that the disadvantages of the extremes of too little and too much room may be perceived. In these cross-bar hives the animal heat rises into the box from the hive, making it as warm as below. The queen often goes up with the bees, and finding it warm and convenient, deposits her eggs, hence young brood as well as honey is found there. We should then be obliged to leave the box on the hive until they hatched, which would make the combs dark, etc.

Boxes set directly on the frames of movable comb hives, will be more speedily filled than when the bees go through the holes, and were it not for the brood, it would always be preferable to put them on in this way, and thus secure the greatest amount of honey.

A BETTER WAY.

Very much may be done to prevent the queen from going into the boxes, by laying on the cross-bars or frames, strips of wood one-fourth inch square, upon which the boxes may be set. They will then be very close to the hive, and the bees will readily find their way into them. If a piece of comb is stuck fast to the bottom of the box, as well as at the top, the bees will commence work a little sooner. Every inducement should be offered to get them into the boxes as soon as possible after the hive is full. Whenever, for want of room, they are compelled to go into another apartment, they will hesitate and lose a little time.

If the honey stored in the frames of the movable comb hive were only in a marketable form, we could get much more in quantity by making our hives large enough to contain a few more frames than would be required to hold winter stores.

ADVANTAGE OF GLASS BOXES.

This advantage attends glass boxes: while being filled, the progress can be watched till they are finished. They should then be taken off to preserve the purity of the combs. Every day that the bees are allowed to pass over them needlessly, renders them darker. Consequently, when our bees are a long time filling a box, the comb is not as purely white as when filled expeditiously. Occasionally, a colony will contain too many bees to work to advantage in one set of boxes. In such a case, after the first are well advanced, raise them up and put another set under them, with holes for communication through both top and bottom.

Two weeks is about the shortest time in which boxes are filled and finished. The time, of course, depends on the yield of honey and size of the swarm. It usually takes three or four weeks.

WHEN TO TAKE THEM OFF.

When no more honey is gathered, all boxes that are worth saving should be taken off. If left longer, the comb not only becomes dark, but unsealed cells containing honey, are often emptied by the bees. The condition of the boxes can be readily ascertained by raising the cover.

If a slide of tin or zinc is used to close the holes when boxes are taken off, some of the bees are apt to be crushed, or find themselves minus a head, leg or wing, and all of them will be irritable for several days. A little smoke will answer every purpose. Raise the box sufficiently to puff under it some smoke, and the bees will leave the vicinity of the holes in an instant. The box can then be removed, and another put on if necessary, without exciting their anger in the least.

Arouse the bees by striking the box lightly four or five times. If all the cells are finished, and honey is still obtained, turn the box bottom up, near the hive from which it was taken, so that the bees can enter it without flying. By this means you can save several young bees that have never left the hive and marked the location; also a few others too weak to fly, which will follow the rest into the hive. Such are lost when we are obliged to take them to a distance. Boxes can be taken off either at morning or evening; if in the morning, they may stand several hours when the sun is not too hot, but on no account let them stand in the sun in the middle of the day, as the combs will melt. The bees will all leave, sometimes in an hour; at others it will take longer. They may be taken off at evening, and let stand until morning, in fair weather; if not too cool, they are generally all out, but when they stand so long, there is some risk of the moth finding them.

HOW TO GET RID OF THE BEES.

When boxes are taken off at the end of the honey season, a different method of getting rid of the bees must be adopted, or we shall lose the honey. Unless the combs are all finished, we shall inevitably lose some, as most of the bees fill themselves before leaving. They carry it home, and return immediately for more, and will take it all if not prevented. It is recommended to take the boxes to some dark room, with a small opening to let the bees out. In the course of the day they will generally all leave, but I have found this method unsafe, as they sometimes find their way back. When a large number of boxes are to be managed, a more expeditious mode is, to have a large box with close joints, or an empty hogshead, or a few barrels with one head out, set in a convenient place. Put the boxes in, one above another, so as not to stop the holes, and throw over the top a thin cloth, to admit the light. . The bees will leave the boxes, creep to the top, and get on the cloth. Turn this over a few times, and you will thus get rid of all the bees with but little loss of honey. All the old bees will return to the hive, but a few young ones will be lost.

BEES NOT DISPOSED TO STING.

Bees seldom offer to sting during this operation, even when the box is taken off without smoke, and carried away from the hive. After a little time, the bees finding themselves away from home, will lose all animosity.

As honey becomes scarce, less brood is reared, and a great many cells become empty, also several cells that contained honey have been drained to mature the portion of brood just started at the time of the failure. We can now understand why our best stocks, that are very heavy, and but a few days before were crowded for room, and storing in boxes, are now eager for honey to store in the hive; as there is room for several pounds. They will quickly remove to the hive the contents of any box left exposed, or even risk their lives by entering a neighboring hive in search of honey.

During a yield of honey, take off boxes as fast as they are filled, and put on empty ones. At the end of the season, take all off. Not one stock in a hundred will starve, that has worked in boxes, that is, when the hive is of the proper size, and was full before adding boxes, unless it has been robbed, or met with some other misfortune.

I prefer taking off all boxes at the end of the first yield of honey, even when I expect to put them on again for buckwheat honey. The bees at this season collect a great abundance of propolis, which they spread over the inside of the boxes as well as the hive; in some instances it is spread on the glass so thickly as to prevent the quality of the honey from being seen. There is no necessity for boxes on a hive, at any time, when there is no yield of honey to fill them. Sometimes, even in a supply of buckwheat honey, a stock may contain too few bees to fill boxes, but just enough to smear them with propolis, which should not be allowed, as it makes them look badly when used another year. At this season, (August) some old stocks may be full of combs, and have but few bees; but when swarms have the hive full in time, they are very sure to have bees enough to work in boxes. I have known them to do so in two weeks after being hived. Some put on boxes at the time the swarms are hived. In such cases the box is often filled first, and will quite often contain brood. I consider it no advantage, but often a damage to do so, unless the swarm is very large and early. I want the hive full in any case, and if they have time to do more, they may then enter the boxes, although they may gather buckwheat instead of clover honey.

When the boxes are free from the bees, two things are to be attended to, if we wish to preserve our honey till cold weather. One is to keep out the worms, the other to prevent souring. The last may be new to many, but it sometimes occurs in warm weather from dampness. The combs become covered with moisture, a portion of the honey becomes thin like water, and turns sour. Remedy: keep perfectly dry and cool, especially *dry*.

TO SECURE HONEY FROM WORMS.

But the worms, you can surely keep them out, you think, since you can seal up the boxes perfectly close, preventing the moth, or even the smallest ant from entering. Yes, you may do this effectually, but the worms will often be there, unless kept in a very low temperature, as in a very cool cellar, or house, and then you have dampness to guard against. I store my surplus honey in a cool, dry cellar, and have no trouble whatever with the moth worm.

I have taken off glass jars, and watched them till the bees were all out, and was *certain the moth did not come near them;* then immediately sealed them up, absolutely preventing any access, and felt quite sure that I should have no trouble with the worms. But I was sadly mistaken. In a few days, I could see a little white dust, like flour, on the side of the combs, and bottom of the jar. As the worms grew larger, this dust was coarser. By looking closely at the combs, a small, white thread-like line could be perceived, enlarging as the worm progressed.

When combs are filled with honey, the worms work only on the surface, eating nothing but the sealing of the cells, seldom penetrating to the centre, unless there is an empty cell. Disgusting as they seem to be, they dislike being daubed with honey. *Wax, not honey, is their food.*

THE WAY THE WORMS GET IN.

The reader would like to know how these worms came in the jars, when to all appearance, it was a *physical impossibility*. I would like to give a positive answer, but can not. I will offer a theory, however, which is original, and therefore open to criticism. If there is any better solution of the problem, I would be glad to hear it.

From the 1st of June till late in the fall, the moth may be found around our hives, active at night, but quiet by day. Her only object, probably, is to find a suitable place to deposit her eggs, where her young may have food. If no proper and convenient place is found, she will be content with such as she *can* find. The eggs *must* be deposited somewhere, and she leaves them in the cracks of the hive, in the dust at the bottom, or outside as near the entrance as she dare approach. The bees running over them may accidentally attach one or more to their feet or bodies, and carry them among the combs where they will be left to hatch. It is not at all probable that the moth ever passed through the hive, among the bees, to deposit her eggs in the jars before mentioned. Had these jars been left on the hive, not a worm would have ever defaced a comb, because, when the bees are numerous, each worm is removed as soon as it commences its work of destruction—that is, when it works on the surface, as it does in the boxes. By taking off these jars, and removing the bees, all the eggs that happened to be there had a fair chance. Many writers finding the combs to be undisturbed when left on the hive till cold weather, recommend *that*, as the only safe way, preferring to have the combs a little darker to the risk of their destruction by the worms. But I object to dark combs; and leaving the boxes on will effectually prevent empty ones from taking their places, thereby involving a loss of surplus honey. I will offer a few more remarks in favor of my view, and then give my remedy for the worms. I have found in all hives from which the bees have been removed in warm weather, say between the middle of June and September, moth eggs enough among the combs to destroy them in a very short time, unless kept in a very cool place. This result has been uniform. Any person doubting this, may remove the bees from a hive full of combs, in July or August. Close it to prevent the entrance of a moth, and set it away in a temperature ranging from 60° to 90°, and if there are not worms enough to satisfy him, he will have better success than I ever did.*

REMEDY.

Whether the foregoing theory is satisfactory or otherwise, we will proceed to the remedy. Perhaps you will find some boxes that will have no worms about them; others may contain from ten to twenty when they have been off a week or more. All the eggs should have time to hatch, which in cool weather may take three weeks. They should be watched, that no worms get large enough to materially injure the combs. Get a close barrel or box that will confine the air as much as possible; in this put the boxes with the holes open. Leave a place for a dish in which to burn some sulphur matches, made by dipping paper or rags in melted sulphur. When all is ready, ignite

---

* Dr. Kirtland, of Cleveland, O., in a lecture before the students of the Medical College, in Cleveland, gave substantially this theory, in accounting for the presence of the worms in the hive. Whether it originated with him or not, he does not say.

the matches, and cover close for several hours. A little care is necessary to use the right quantity ; if there is too little, the worms are not killed ; if too much, it gives the combs a green color. A little experience will soon enable you to judge. If the worms are not killed on the first trial, another dose must be administered. Whether this gas from burning sulphur will destroy the eggs of the moth, I have not tested sufficiently to decide ; but I do know that it is an effectual quietus for the larvæ.

Much less sulphur will adhere to paper or rags, when it is very hot, than when just above the temperature necessary to melt it. This should be considered, also the number of boxes to be treated, size of the barrel used, etc.

Boxes taken off at the end of warm weather, and exposed to cold through the winter, will have all the worms as well as eggs destroyed, consequently boxes so exposed may be kept any length of time, if the moth is carefully excluded.

CHAPTER X.

SWARMING.

The subject now before us is one of exciting interest. The prospect of an increase of stocks is sufficient to engage the attention of the apiarian, even when the phenomenon of swarming would fail to awaken it. But to the naturalist this season has charms that the indifferent beholder can never realize.

KNOWLEDGE NECESSARY.

It is important that the practical apiarian, as well as the naturalist, should have a thorough knowledge of this branch of apiculture, and not accept any assertions without evidence. Twelve years ago, I found it necessary to establish many positions with facts, and also to give the manner of obtaining them. But I now have the movable comb hive, which gives ocular demonstration of what then appeared to be mere conjecture, and it will be unnecessary to specify in every case the process by which I have arrived at certain conclusions. I trust that the objector will see the necessity of depending upon facts, instead of any notion, imbibed from nursery tales. Neither will it always do to reason analogically, because nature nowhere gives us an exact parallel.

A noted politician who has reached an eminent position as a legislator, declared a short time since, that the queen bee was a myth—that she existed only in the imagination of ignorant bee-keepers. Every man who has taken the first step in the investigation of apiarian science, knows that he made a fool of himself quite unnecessarily.

WHEN SWARMING COMMENCES.

The swarming season in this latitude sometimes commences May 15th, and at other times, July 1st. It usually ends about the 15th of the latter month. I have known two seasons in Montgomery Co., N. Y., when swarms continued to issue throughout the entire summer, beginning in May and ending August 25th, with no interval of more than a week without swarms. One of these, 1863, was wet, and the flowers yielded but little honey. The native bees sent out about one-third the usual number of swarms, while the Italians continued to swarm for three months. They did not store much more honey than others, but they must have collected more to feed the greater quantities of brood which they reared. Rather than be idle when the yield was scanty, they collected material, made combs, reared brood, and sent out swarms; and at the end of the season the colonies were as strong, and had as much honey as the natives, which had not swarmed.

The bee-keeper who thinks much of his bees, will, of course, wish to see and hive his swarms as they issue. If neglected, for even a short time after they cluster, they will often relieve themselves from such protection, and seek the shelter of some old tree in the woods, humming indignant reproaches as they leave. Without some knowledge of the indications of swarms, we often watch vainly for weeks, remaining at home, and perhaps neglecting important business in the fruitless expectation that the "bees will swarm." External appearances are not to be depended on. Very early swarms often issue before clustering out; also, they very often cluster out without swarming. It is necessary to look inside the hive for reliable indications.

I have several apiaries away from home that need attention in the swarming season, and a person must be on hand each fair day, to take charge of the swarms as they issue. To avoid watching unnecessarily before they begin, and after they cease, some one examines the hives in the middle of the day.

INDICATIONS.

If they are box hives, a little smoke is blown under, the hives turned over, the bees driven away with a little more smoke, and the queen cells examined. If there are none containing eggs or larvæ, or none with thin smooth walls, evidently just begun, there is not much prospect of a swarm for several days. There is always a possibility, however, that some cells will be out of sight. But if any cells contain eggs, or larvæ nearly ready to seal over, or actually sealed over, we know at once what to expect. When sealed over, the swarm will probably issue the next day. This is at the beginning of the season. If at, or near the close, we examine again, and find the sealed queens destroyed, we at once conclude that they are done swarming.

CARE IN EXAMINING BEES.

Full hives require a little more care in turning over, than others. You need not be in fear of the bees, running up the sides of the hive; they will not sting. Lift the hive carefully, and avoid breathing among them, except to blow the smoke. It can be done at morning or evening, but more bees are in the way, and they are more inclined to be cross. In operating with the native bees, protection for the face or hands is hardly necessary, but with the Italians it would sometimes be well to put something over the face.

To a person who has never inverted a hive full to overflowing of bees, or has never seen it done, it appears like a great undertaking, as well as the probable ruin of the stocks. But after the first trial, the magnitude of the performance is greatly diminished, and will grow less with every repetition of the feat, until there is not the least dread attending it. Without smoke, I hardly deem it practicable, but with it there is not the least difficulty. It would be very unsatisfactory to turn over a hive, and have nothing with which to drive the bees away from the very places which you wish particularly to inspect. The smoke is just the thing to do it.

I never discovered any bad effects of such overturning and smoking.

With the movable comb hive we have only to lift out the frames, to be able to predict when a swarm may be expected.

I have found the requisites for all regular swarms to be something like this. The combs must be crowded with bees; they must contain a numerous brood advancing from the egg to maturity, and the bees must be obtaining honey either from flowers, or artificial sources.* A surplus of bees in a scarcity of honey is insufficient to bring out the swarm, neither will plenty of honey suffice, without the bees and brood. The period of proper duration in which all these conditions exist, will vary in different stocks, and many times does not occur at all during the season.

These causes appear to result in the construction of queen cells, generally begun before the hive is filled; sometimes when only half full.

PREPARATIONS FOR SWARMING.

They are about one third done when they receive the eggs; as these eggs hatch into larvæ, others are begun, and receive eggs at different periods for several days later. The number of such cells seems to be governed by the prosperity of the bees; when the family is large, and the yield of honey abundant, they may construct twenty, at other times not more than two or three, although several such cells may remain empty.

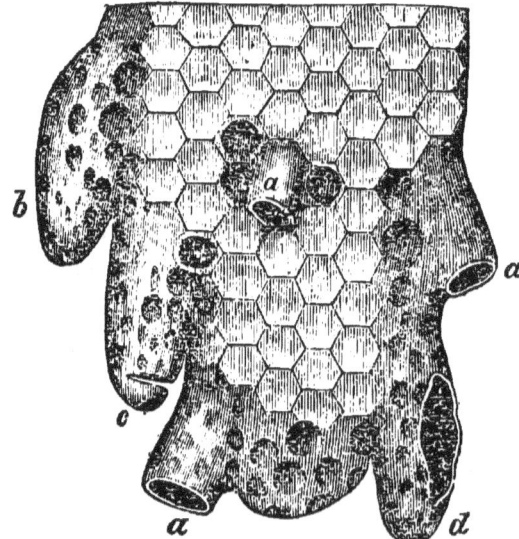

Fig. 22.—CLUSTER OF QUEEN CELLS.
a, a, a, Size of the cell when the egg is deposited; *b,* Finished cell; *c,* Cell from which a mature queen has emerged; *d,* Cell in which the queen has been destroyed by a rival, and removed by the workers.

I have already said that a failure, or even a partial one,

---

* The Italians will swarm sooner than the native bee when obtaining but little honey.

in the yield of honey at any time after the royal eggs are deposited, before the sealing of the cells—which is about ten days—will be likely to ensure their destruction. Even after being sealed, I have sometimes known them to be destroyed.

WHEN SWARMS ISSUE.

But when there is nothing precarious about the supply of honey, the sealing of these cells indicates the first swarm, which will generally issue on the first fair day after one or more are finished. I have never missed a prediction of a swarm, when I have judged from these signs, in a prosperous season.

When there is a partial failure of honey, the swarm will sometimes wait several days after finishing these cells. If the family is strong, and there is a sudden increase in the yield of honey, the swarm may not wait for the sealing of any cells, but will issue about the time, or very soon after, eggs are laid in them. This occurs sufficiently often to be mentioned as an exception to the rule. But nevertheless, the rule is, to expect the first swarm upon the sealing of any of the royal cells.

Again, if there is a failure of honey when these cells are finished, there may be no swarm. A failure often occurs between fruit blossoms and white clover, and also at the end of the honey season, whether it be the first, middle, or last of July. The first deficiency occurs about June 1st. If fruit blossoms have yielded only a moderate amount of honey, the strong stocks that have a good supply, feeling their importance, like some specimens of the human family, who consider a moderate competence inexhaustible, will indulge in extravagance by rearing a useless number of drones. When the income ceases, and famine is close at hand, something must be done to save the colony. The drones are sacrificed for the good of the community; even the brood is destroyed, queen cells demolished, and all idea of swarming given up. The destruction of drones at any time may be accepted as evidence that, for the time being, swarming is over. Colonies possessing but a moderate supply of bees and honey, usually work on safe principles; they can not afford to rear any drones, and when the scarcity between fruit blossoms and clover occurs, they pass the crisis without any sacrifice, and are ready to take advantage of the first yield, and will throw out swarms long before those who were apparently far more prosperous in the beginning of spring. This explains how a second-rate stock may sometimes surpass a "No. 1" in swarming, which has been quite a mystery to many bee-keepers.

WHY DRONES ARE SOMETIMES KILLED IN SPRING.

I have seen statements going the rounds of the agricultural press like this: "There will be no swarms this year, as the bees have killed off their drones." This fact does not settle the matter for the season, by any means, but it will assuredly be some weeks before they can possibly get another brood of drones under way. After a reverse of this kind, they will not begin again until honey is obtained in abundance, and it is quite often that all the conditions are not present again, until the season is so far advanced that it is too late. Occasionally they make preparations the second time, and again abandon them. It is quite unusual for none of them to send out swarms late in the season.

Who will say that bees do not manifest wisdom? What prudent man would emigrate with a family if famine were plainly indicated, when by remaining at home, he would have a present abundance? Who can fail to admire this wise and beautiful arrangement? The combs must contain brood; the bees must find honey during the rearing of the queens. If a swarm were to issue as soon as honey were obtained, the consequences might be fatal, as there would not be a numerous brood to hatch out, and replenish the old stock with bees enough to keep out the worms. Were they to issue at any time, as soon as the bees had increased enough to spare a swarm, without regard to the yield of honey, they might starve.

WHICH BEES ISSUE.

I find many theories conflicting with these views which need attention. It is generally supposed that a young queen must be matured to issue with the swarms, and that the old queen and old bees are permanent residents of the old hive. It is probable that no rule governs the issue of the workers. Old and young come out promiscuously. That old bees issue with a swarm is evident from the fact that sometimes not a quarter as many will be left as commenced work in the spring. Also a great many may be seen in late swarms, with wings so worn as to be unable to fly with the load of honey which they attempt to carry. I have seen enough get down in this way, from one swarm, to fill a pint measure.

That young bees leave, any one may be satisfied on seeing a swarm issue. A great many too young and weak to fly, will drop down in front of the hive, having come out now for the first time, perhaps not an hour out of the cell; these very young bees may be known by their color. That these may creep back to the hive, is another inducement to set it near the ground.

The old queen often gets down in the same way, but her burden of eggs is probably the cause of her inability to fly.

THE OLD QUEEN LEAVES.

That the *old* queen leaves with the first swarm is so easily proved with the movable comb hive, that it is un-

7*

necessary to occupy several pages in maintaining it. After the swarm has left, you have only to examine the combs, to be assured that she is nowhere in the hive. The absence of eggs in the cells is other proof.

HIVES SHOULD BE READY.

We will now suppose that some of your colonies are ready to send out swarms, and will also presume that the empty hives for the reception of swarms are in readiness. To prepare a hive after the swarm has issued, indicates bad management; negligence here, argues negligence elsewhere; it is one of the premonitions of "bad luck."

You will also want a number of bottom boards expressly for hiving. Get a board a little larger than the bottom of the hive, nail strips across the ends on the under side, to prevent warping; in the middle cut out a space five or six inches square, and cover with wire-cloth. These are for your large swarms in very hot weather, to be used for four or five days. It is much safer to use them than to raise the hive an inch or two for ventilation. They are also essential on many other occasions.

IMMEDIATE INDICATIONS OF A SWARM.

When the day is fair, and there is not too much wind, first swarms generally issue from 10 A. M., till 3 P. M. The first outside indication of a swarm will be an unusual number of bees about the entrance, from one to sixty minutes before they start. The utmost confusion seems to prevail, bees run about in every direction, and the entrance is apparently closed by the mass of bees; presently a column from the interior, forces a passage to the open air; they rush out by hundreds, vibrating their wings, and when a few inches from the entrance, rise in the air; some rush up the side of the hive, others to the edge of the bottom-board. If you have seen the old queen come out .

the first one, and the rest following her, as we are often told she does, you have seen what I never did in a first swarm. I have occasionally seen the old queen issue, but not before the swarm was half out. Second and third swarms conduct themselves quite differently.

The bees, when first rising from the hive, describe circles of but few feet, but as they recede, they spread over an area of several rods. Their movements are much slower than usual. In a few moments thousands may be seen revolving in every possible direction. A swarm may be seen and heard at a distance where fifty hives at ordinary work would not be noticed.

SWARM CLUSTERS.

When all are out of the hive, or soon after, some branch of a tree or bush is usually selected upon which to cluster. In less than half a minute after the spot is indicated, even when the bees are spread over an acre, they are gathered in the immediate vicinity, and all cluster in a body, in from five to ten minutes after leaving the hive. They should be hived immediately, as they show impatience if left long, especially in the sun; also if another colony should send out a swarm while they were hanging there, they would be quite sure to unite.

HOW TO DO IT.

It makes but little difference in what way they are put into the hive, provided they are all made to go in. Proceed as is most convenient; an old table or bench is very good to keep them out of the grass, should there happen to be any. If there is nothing in the way, lay your bottom-board on the ground, make it level, set your hive on it and raise one edge an inch or more with small sticks or stones, to give the bees a chance to enter.

Cut off the branch on which the bees are hanging, if it

can be done as well as not, and shake them off in front of the hive; a portion will discover it and will at once commence a vibration of their wings, which seems to be a call for the others. A knowledge that a new home is found, seems to be communicated in this way, as it is continued until all have entered. A great many are apt to stop about the entrance, thereby nearly or quite closing it, and preventing others from going in. You can expedite their progress by gently disturbing them with a stick or quill. When gentle means will not induce them to enter in a reasonable time, and they appear obstinate, a little water sprinkled on them will greatly facilitate operations. Be careful and not overdo the matter by using too much water; they can be made so wet that they will not move at all.

When they cluster on a branch that you do not wish to cut off, arrange the bottom-board as before directed, then turn the hive bottom up directly under the main part of the cluster, and if you have an assistant, let him jar the branch sufficiently to dislodge the bees; most of them will fall directly into the hive. If no assistant is at hand, strike the under side of the branch with the bottom of the hive, and when the bees have fallen in it, set it on the board; the sticks will prevent the bottom from crushing the bees.

I have gone up a ladder twenty feet high, got the bees in the hive in this way, and backed down without difficulty. After putting the hive in its place, sometimes a part of the bees will go back to the alighting place; in that case a small leafy branch should be held directly under and close to them, and as many jarred on it as possible. Hold this still, and shake the other to prevent their clustering there; you will soon have them all collected, ready to bring down and put by the hive.

A basket or large tin pan may be taken up the ladder instead of the hive, from which the bees can be readily emptied before it. But very few will fly out in coming down. If you succeed in getting nearly all the bees at the first effort, merely shaking the branch will be sufficient to prevent the remainder from alighting, and will turn their attention below, where those which have already found a hive will be doing their best to call them. When the hive is first turned over, most of the bees will fall on the board and rush out, but as soon as they realize that a home is found, they will commence buzzing. This quickly communicates the fact to those outside, who immediately turn about and hum in concert while marching in.

Another plan may be adopted if they light very high, when the branch is not too large, and there is not too much in the way below it. Have ready two or three light poles of suitable length, with a branch at the upper end, large enough to support a bushel basket. Raise the basket directly under the swarm, and with another pole dislodge all the bees. They will fall into it and may be quickly let down. Now, if you have secured nearly all, throw a sheet over them for a few moments to prevent their escape. They soon become quiet, when they may be hived, and but few will return to the branch, as many will do when they are put in the hive immediately. When many swarms are to be brought down in this way, a bag may be prepared which can be put up among the branches where it would be difficult to use a basket. Sew a hoop around the top of a bag, and fasten one side of it to a piece of wood two inches square, and three feet long, four inches from one end. Brace the hoop with a strong wire fastened six inches below. Around the lower end of the stick, fasten a band of hoop iron twelve inches long, in such a manner that it will form a loop two inches in diameter, on one side, and make another loop one inch in diameter, in the same way, one foot above. Light poles of different lengths should be fitted into these bands.

When ready to operate, put in a pole long enough to

reach the bees, and raise the mouth of the bag directly under them. With the end that projects above the hoop, strike the limb upon which the bees have clustered, with force sufficient to jar them into the bag. By dexterously

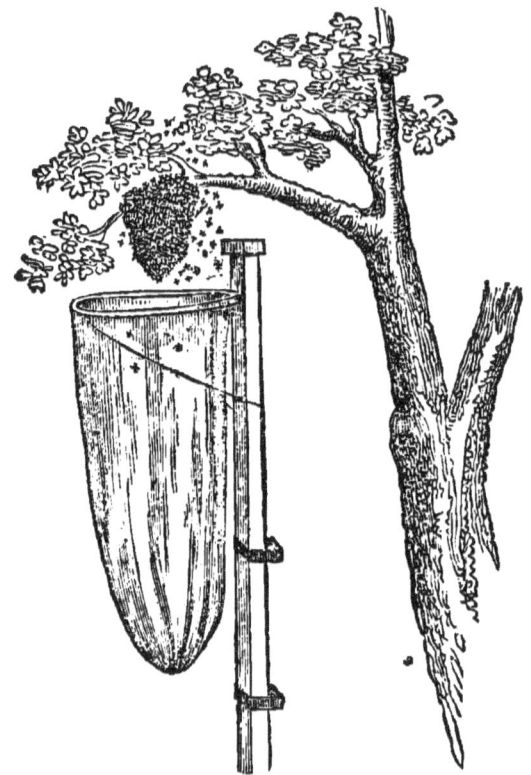

Fig. 23.—BAG FOR HIVING BEES.

tipping it sideways, the mouth of the bag can be effectually closed, and the bees brought to the ground without difficulty. In a few minutes they become quiet, and can be hived as before mentioned.

This method is generally to be preferred to ascending the ladder.

Bees often begin to cluster near the ground, in a convenient situation for hiving. In such a case I do not wait for all to alight, but as soon as such place is indicated, I get the board and hive ready. When a quart or so are gathered, I shake them in the hive, and set it up; the swarm will now go to that, instead of the branch, especially if the latter is shaken a little. Where many bees are kept, it is advisable to be as expeditious as possible. A swarm will thus be hived much sooner than when allowed to cluster.

Swarms will sometimes alight in places where it is impossible to jar them off, such as a large limb, or trunk of a tree. In which case place the hive near, as first directed; take a large tin dipper—the most convenient vessel for the purpose—and dip it full of bees; with one hand turn back the hive, and with the other put the bees into it.* Some of them will discover that a home is provided, and set up the buzzing. The remainder can be emptied in front of the hive as you dip them off. I have known a few instances when the first dipper full all ran out and joined the others without making the discovery that they were in a hive, but this is seldom the case. When you get the queen in, there will be no trouble with the remainder, even if there are many left.

As soon as they ascertain that the queen is not among them, they will manifest it by their uneasy movements.

ALL SHOULD BE MADE TO ENTER.

They will soon leave and join those in the hive, or if the queen is yet on the tree, even if there be but a dozen with her, those in the hive will leave and cluster again. In all cases they must all be made to enter; a cluster outside may contain the queen, unconscious of a home, and the consequence might be, her departure for a miserable one in the woods.

---

* Dipping is preferable to brushing with a wing or broom, as the latter irritates them exceedingly.

CARRY TO THE STAND.

When all are in, except a few that will be flying, let the hive close down to the board, take hold of this and carry it at once to the stand which they are to occupy, and raise the front edge half an inch, unless you use the bottom board recommended. Let the back rest on the board, that they may have means to re-ascend, if they chance to fall, which large swarms often do in hot weather. If the bottom is an inch or more from the board when they thus fall, there is nothing to prevent their rushing out on every side; they can not easily get up again, and if the queen comes out with the rush, there are some chances of their leaving.

TO PUT THEM IN A MOVABLE COMB HIVE.

There are but few movable comb hives which the bees will as readily enter as the box. It is usually the least trouble to put the swarm in a box hive, and transfer to the other, near evening, or at your first leisure. A plain box made of thin boards, is lighter and easier handled than a common box hive. Carry the swarm to the stand, and if the hive is like the one I use, remove the honey board, spread the frames each way from the middle, and shake the bees directly in the hive between the frames, as you would a quantity of grain. Before many can creep out, lay on the honey board. As the hive is close to the bottom, no bees can get out except at the entrance, and these will immediately turn about. If a few are yet adhering to the box, give it a jar in front of the hive. After all are quiet, the frames may be properly adjusted.

SHADE IMPORTANT.

It is very important that swarms should be protected from the sun for several days in hot weather, from nine o'clock till three or four; and then if the heat is very oppressive, and the bees cluster outside, sprinkle them with water and drive them in. Wetting the hive occasionally will carry off a large portion of the heat, and make it much more comfortable.

CLUSTERING BUSHES.

If there are no large trees in the vicinity of your apiary, all the better; there will then be no trouble with the swarms lighting out of reach, but all bee-keepers are not so fortunate. In a place where there are no natural conveniences, it is necessary to provide something for them to cluster on. Get some bushes six or eight feet high— evergreens are preferable—cut off the ends of the branches except a few near the top, secure the whole with strings, to prevent swaying in ordinary winds, make a hole in the earth deep enough to hold them, and so large that they may be easily lifted out. The bees will be likely to cluster on some of these; they can then be raised out and the swarm hived without difficulty. A bunch of dry mullen-tops tied together on the end of a pole, makes a very good place for clustering; it so nearly resembles a swarm that bees themselves appear to be sometimes deceived. I have frequently known them to leave a branch where they had begun to cluster, and settle on this when held near.

The reasons for immediately removing the swarm to the stand, are, that they are generally more convenient to watch in case they are disposed to leave, and many bees can be saved. All that leave the hive, mark the location the same as in spring. Several hundreds will probably leave the first day, a few, several times. When removed at night to the permanent stand, such will return to the stand of the previous day, and are generally lost, whereas, if they are removed at once, this loss is avoided.

Those that are left flying at the time, return to the old stock, which those that return from the swarm the next day will not always do. It will take no longer to move them at one time than at another. It is useless to object, and say that "it will take too long to wait for the bees to get in." I shall insist on your making all the bees enter before you leave them. I consider this an essential feature in the management. I will not say that my directions will *always* prevent their going to the woods, but in my experience, not one in a thousand has ever thus left. It is possible that judicious management has had no influence upon my success, yet I have indulged something like an opinion of this kind for a long time.

LOSS BY FLIGHT.

Some of my neighboring bee-keepers lose a quarter or half their swarms by flight, and how do they manage? When the word is given out, "Bees swarming," a tin horn, tin pan, or any thing to make a horrible din, is seized upon, and as much noise made as possible, to *make* them cluster, which they naturally would do, without the music. The fact that they would cluster in any case, probably gave rise to the opinion of the old lady who *knew* "drumming on a tin pan did good, for she had tried it." Very often a hive is to be constructed, or an old one, unfit to use in any shape, must have some new cross sticks; or something else must be done to take time. When the hive is obtained, it must be washed with something nice to make the bees like it; a little honey, or sugar and water, molasses and water, salt and water, must be daubed on the inside; or salt and water rubbed on with hickory leaves is "the best thing in the world;" several other things are just as good, and some are better. Even whiskey, that bane of man, has been offered them as a bribe to stay, and sometimes they endure these nuisances, and go to work.

NOTHING BUT BEES NECESSARY IN A HIVE.

I will not say positively that all these things do harm, yet I am quite sure that they do no good, as nothing is needed but bees in a hive. Is it reasonable to suppose that they are fond of all the "knick-knacks" given them? *I have never used any*, and could not possibly have done much better. I am careful to have the hive sweet and clean, and not too smooth inside; an old hive that has been used before, is scalded and scraped.

But to the manner in which people get the bees in, after the hive is ready. A table with a cloth spread over it, is set out, and the hive prepared as above, is set upon it. If they succeed in getting the swarm even on the outside of the hive, it is left; if it goes in, it is well; if it goes off, "better luck next time." The hive is left unsheltered in the hot sun, and when there is no wind the heat is soon insupportable, or at least very oppressive. The bees hang in loose strings instead of a compact body, as when kept cool. They are very apt to fall, and when they do, will rush out on every side; if the queen chance to drop with them, they *may* "step out." Two-thirds of all the bees that go to the woods, are managed in this, or a similar manner, and may it not be said they are fairly driven off?

Hives painted some dark color will become intolerably warm in the sun, and are often deserted. The rank smell of newly painted hives of any color, often causes the bees to leave for more pleasant quarters in the woods.

Perhaps one swarm in three hundred will depart for the woods without clustering. But I have never had one leave me thus. Yet I have indisputable evidence that some will do it.

DO THEY SELECT A HOME BEFORE SWARMING.

The inquiry is often made, do all swarms have a place selected before leaving the parent stock? The answer to this must ever be conjectural. I could relate some circumstances favoring the affirmative, and as many for the negative, but will let it pass. Yet I think if bees are properly cared for, that ninety-nine swarms in a hundred will prefer a good clean hive to a rotten tree in the woods.

HOW FAR WILL THEY GO.

How far they will go in search of a home is also uncertain. I have heard of their going seven miles, but could not learn how the fact was proved. I have no experience of my own upon this point, but will relate a circumstance that happened near me. While a neighbor was plowing, a swarm passed over him; being near the earth, he "pelted them heartily" with loose dirt, which brought them down, and they clustered on a low bush; they were hived and gave no further trouble. A man living some three miles from this one, on that day hived a swarm about eleven o'clock, and left them to warm up in the sun, as just described. About three o'clock, their stock of patience being probably exhausted, they resolved to seek a better shelter. They departed in a great hurry, not even waiting to thank their owner for the spread on his table, and the choice perfumery with which he had scented their hive. They gave him no notice whatever of their intention "to quit," until they were moving! With all their goods ready packed, they were soon under way, accompanied by their owner with music, but whether they marched with military precision, is uncertain. In this case, the bees took the lead; the man with his tin-pan music, kept the rear, and was soon at a respectful distance. They were either not in a mood, just then, to be charmed by melodious sounds, or their business was too urgent to allow them to stop and listen. Their means of locomotion being superior to his, he gave up in despair and out of breath, at the end of a mile.

Another person, about the same time in the day, saw a swarm moving in the same direction; he also followed them till compelled to yield to their greater locomotive powers. A third discovered their flight, and likewise attempted a race, but like the others, was soon left behind. The before-mentioned neighbor saw them, and stopped them as described.

How much farther they would have gone, it is of course impossible to say. That it was the same swarm that started three miles away, appears almost conclusive.

We will now return to the issuing of the swarms. There will be some emergencies to provide for, and some exceptions to notice.

If we keep many colonies, the chances are that two or more may issue at one time; and when they do, they will nearly always cluster together. It is plain that the greater the number of colonies, the more such chances are multiplied.

ONE FIRST SWARM HAS BEES ENOUGH.

One first swarm, if of the usual size, will contain bees enough for profit, yet two such will work together without quarreling, and will store about one-third more than either would alone; that is, if each single swarm would gather fifty pounds, the two together would not get over seventy pounds, perhaps less. Here then is a loss of thirty pounds, besides the virtual loss of one of the swarms for another year; because such double stocks are not generally any better the next spring, and not often as good as single ones. Hence the advantage of keeping the first swarms separate, is apparent.

HOW TO KEEP SWARMS SEPARATE.

"Prevention is better than cure." We can, if we are watchful, often prevent the issue of more than one at a time. This depends in a great measure, on our knowledge of indications. I have said that before beginning to fly off, they were about the entrance in great numbers; there may be one exception in twenty, when the first indications will be a column of bees rushing from the hive. To pursue our investigations a little further, we will look within, that is, if glass boxes are used, such as have been recommended. It is an advantage to know which are about to cast swarms, as long beforehand as possible.

These glass boxes are usually filled with bees; previous to leaving they may often be seen in commotion long before any unusual stir is visible outside, sometimes for nearly an hour. The same may be noticed in a glass hive. In good weather, when we have reason to expect many swarms, it is our duty to watch closely, especially when the weather has been unfavorable for several days previous. A number of colonies may have finished their queen cells during the bad weather, and be ready to send out swarms within the first hour of sunshine that occurs in the middle of the day. We must expect this to take place sometimes, and in large apiaries there is apt to be trouble, unless proper precautions are taken. It is well to know by previous examination, which hives have made preparations for swarming, and as soon as one has begun to issue, look at all the rest that are in condition to swarm; or, what is much better, look before any have started. Even if nothing unusual is perceived about the entrance, raise the cover to the boxes. If the bees there are all quiet as usual, no swarm need be immediately apprehended, and you will probably have time to hive one or two without interruption.

But should you discover the bees running to and fro in great commotion, although quiet at the entrance, you should lose no time in sprinkling those outside with water. They will instantly enter the hive to avoid the apprehended shower. In half an hour they will be ready to start again, during which time the others may be secured. I have had, in one apiary, sixteen hives all ready in one day, all of which actually swarmed, and several would have started at once had they not been kept back, allowing only one to issue at a time, as described. They had been hindered by the clouds, which broke away about noon.

When any of the subsequent swarms were disposed to unite with those already hived, a sheet was thrown over to keep them out. I had four so covered at once. An assistant is very useful at such times; one can watch symptoms, and detain the swarms, while another hives them. Occasionally when waiting for a swarm to start, two may do so simultaneously.

CAN NOT BE STOPPED WHEN PART ARE ON THE WING.

Whenever a part was already on the wing, I never succeeded in retarding the issue; it is then useless to try to drive or coax them back. To succeed, the means must be applied before any part of the swarm leaves.

Two or more swarms will cluster together and not quarrel, if put into one hive. I have already mentioned the disadvantages. Unless business is very urgent, your time can not be better employed than in dividing them

HOW TO DIVIDE.

But it is necessary first to provide a stock of patience, as it may be a long job. Spread a sheet on the ground, shake the bees upon the centre of it, and set an empty hive each side of the mass, with the edges raised to allow the bees to enter; if too many are disposed to enter one hive, set it farther off. If they cluster in a situation where they can not be hived in a body, they may be dipped off as before directed, but instead of putting them all in one hive, put a dipper full in each, alternately, till all are in. They should be made to enter rapidly; keep the entrance clear and stir them up often, or sprinkle a very little water on them, as they should not be allowed to stop their humming until all are inside. There are even chances of getting a queen in each hive. The two hives should now be placed twenty feet apart; if each has a queen, the bees will remain quiet, and the work is done; but if not, the bees in the one destitute will soon manifest it, by running about in all directions, and when the queen can not be found, will leave for the other hive where there are doubtless two, a few going at a time. There are two or three methods of separating these queens. One is, to empty the bees out, and proceed as before, a game of chance that may succeed at the next trial, and may have to be indefinitely repeated. Or, as soon as it is ascertained which is without a queen, spread down a sheet, set the hive on it, and tie the corners over the top to secure the bees. Turn the hive on its side for the present, to give them air, or let it down on a wire cloth bottom-board, and stop the hole in the side. The bees would be less likely to be smothered if the hive could be secured to the bottom board and lie on its side. When these are secured, get another hive, and jar out those with the queens. Let them enter as before, and then set them apart, watching the result. If the queens are not yet separated, it will soon be shown. The process must be continued till successful, or the bees with the queens may be easily looked over, and one of them found. Indeed, a sharp look out should be continued from the beginning, and one of the queens caught if possible.

No danger of her sting need be apprehended, for she will not demean herself to use it against a plebeian foe—she must have a royal antagonist. When successful in obtaining one, put her in a tumbler or some safe place;

then put the bees in two hives, place them as directed, and you will soon learn where your queen is needed. After the work is completed, the hives should be at least twenty feet apart; perhaps forty would be still better.

When two swarms are mixed, and then separated, it is evident that a portion of each swarm must be in both hives. The queen in each must be a stranger to a part of her subjects; these might, if their own mother was too near, discover her, and leave the stranger for an old acquaintance, and in the act, attract the rest with them, including the queen. I have known a few instances of the kind. If you are disposed to separate them, but are afraid to work among them to this extent in the middle of the day, or if there is danger of more issues to mix with them, and add to the perplexity of which you already have enough, then you can hive them as a single swarm, but instead of using a bottom board, invert an empty hive, and set the one containing the swarm, on it, and insert a wedge between them on one side, for ventilation. Many bees are liable to drop down, but the lower hive will catch them, and there is less danger of their leaving.

Let them remain till near sunset, when another course must be taken to find a queen, though by that time one is sometimes killed—yet it is well to know the fact. Take them to some place out of the sun, as a less number will fly during the operation.

Look in the lower hive for a dead queen, and if you find none, look thoroughly as far as possible, for a compact cluster of bees, the size of a hen's egg, that may be rolled about without separating. Secure this cluster in a tumbler; it is quite likely that one of the queens is a prisoner in the middle.* Should two be seen, secure both. Then

---

* All stranger queens, introduced into a stock or swarm, are secured and detained in this manner by the workers, but whether they dispatch them, or this is a means adopted to incite them to a deadly conflict, writers do not agree, and I

divide the bees and give the destitute one a queen; or, if you have caught two, one to each. It would be well first to see if the queen is alive, by removing the bees from about her. But should you find no cluster of the kind, spread a sheet on the ground, shake the bees on one end of it, and let them march towards the hive at the other end. You may now see the cluster, and may not, but they will spread out, and give you a good opportunity to see her majesty. When you discover her, secure her by setting a tumbler over her. If there are a few bees shut up with her, there is no harm done. Slip a piece of window-glass under, and you will have her safe, and by this time you will know what is to be done next. This operation could not well be performed in the middle of the day, or in the sun, as so many bees would be flying that they would greatly interfere.

Should you fail to find a queen, and be unable to make a division in consequence, or resolve from want of time, patience, or energy, to let them remain together, it is unnecessary to put them in any larger hive than usual; they will certainly have room enough by cold weather. If there are more than two together, they should be divided by all means. When two large swarms are left together, it is necessary to keep an inverted hive under them for the first three or four days, but no longer, as they might extend their combs into the lower hive. When the lower hive is removed, boxes should be immediately put on, which should be changed for empty ones, as fast as they are filled. Yet this extra honey is of not quite as much advantage as an increase of stocks; when the latter is an object, I would recommend the following disposition of the swarms.

can not say, as I never saw the bees voluntarily release a queen thus confined. But I have seen queens, when not prevented by the bees, rush together in a fatal encounter, of which one was soon left a fallen victim. It is said that it *never* happens that both are killed in these battles—perhaps not. I never saw *all* of these royal combats, and, of course, am not competent to decide.

Return one-third or more of them, without any queen, to one of the old stocks. They will immediately enter without any contention, and issue again in about nine days, or as soon as a young queen is matured to go with them. There may be exceptional cases. I would recommend this course in all cases of the kind, but they are apt to be rather idle, even when they might labor in the boxes, and there is often a loss of some eight or ten days. The collections of a good swarm may be estimated at from one to three pounds per day. A swarm that just fills the hive, would gather, from ten to twenty-five pounds of box honey, if it could have been located ten days earlier. Still another plan may be adopted, when you have a very small swarm that is not likely to fill the hive, and has not been hived more than two or three days.

Put one-third of your two swarms in with that, taking care, as before, not to let your only queen go with them.

The manner of doing it is very simple. Put them in a hive as before directed, and jar them out in front of the one you wish them to enter, or invert it, setting the other over, and let them go up.

Except on the day of swarming, care is necessary not to introduce a small number with a large swarm, as they are liable to be destroyed. The danger is much greater than to put equal numbers together, or a large number with a few. On the day that swarms issue, they will generally mix peaceably, but in proportion as time intervenes between the issues, the liability to quarrel will increase. Yet, I have united two families of about equal numbers in the fall and spring, and with a few exceptions, have had no difficulty.

DIFFERENT PROCESS WITH MOVABLE COMB HIVE.

The foregoing remarks are for those who use the box hive. But those using the movable combs will have less trouble. All the bees may be put into one hive, with the surplus boxes on the top, and if necessary an empty hive under. There is room on the top for one-third more boxes, than on the box hive. If the weather has been good, they may be divided in a week, by putting half the combs in an empty hive, and proceeding as directed in chap. XI.

If you have empty combs on hand, divide the swarms at once. As soon as you ascertain which has *no* queen, shut it up, and when it is so dark that the bees will not fly, put them in the hive containing combs. Previously insert a small piece of comb containing brood, from which they may raise a queen, that is, when you can not furnish a queen or queen cell.

If you have a laying queen to spare, it is not all important to have combs for the queenless division; simply put them in an empty hive, and give them a caged queen. If they do not stay willingly, confine them a day or two; when the queen is set at liberty they will usually be contented.

The difference in time gained by giving them a laying queen, instead of the means of rearing one, is about three weeks, equivalent in value to a small swarm. I have but little doubt that an improved system of bee culture will make it profitable to rear queens, and keep them on hand for such emergencies, as well as for all occasions where new queens are needed. It will sometimes insure a gain equal to the difference between a fair profit and actual loss. Should a new swarm lose its queen, you may introduce one in a cage immediately, liberating her in about forty-eight hours.

Another means of keeping swarms separate is the "swarm-catcher," made by covering a frame with fine netting, to be set before the hive when the swarm is issuing. But as it seems to keep back part of the swarm, and is also open to other objections, I have laid it aside.

SWARMING.

SWARMS SOMETIMES RETURN.

Occasionally a swarm will issue, and in a few minutes return to the old stock. The most common cause is the inability of the old queen to fly, on account of her burden of eggs, or old age. I have sometimes, after the swarm had returned, found the queen near the hive, and put her back; and the next day she would come out again, and fly without difficulty, probably having discharged some of her eggs.

They are more apt to return in windy weather, or when the sun is partially obscured by clouds. About three-fourths of such swarms will not re-issue until a young queen is matured, eight or ten days afterwards, and a few not at all. But when the queen returns with the swarm, they usually come out again the next day, but sometimes not before the third or fourth day after. I have known a few instances, when they issued again the same day.

Sometimes a swarm will issue and return three or four days in succession, but this may generally be remedied, as it is often owing to some inability of the queen; and she may frequently be found while the swarm is leaving, outside the hive, unable to fly. In such circumstances, have a tumbler ready and secure her as soon as she appears. Get the empty hive for the swarm, and a large cloth, and put down a bottom-board a few feet from the stock. The swarm is sure to come back, and the first bees that alight on the hive will set up the call. As soon as you perceive this, lose no time in setting the old stock on the board at one side, throwing the cloth over it to keep out the bees. Put the new one in its place on the stand, and the queen in it; in a few minutes the swarm will be in the new hive, when it can be removed and the old one replaced. But should the swarm begin to cluster in a convenient place, when you have so caught the queen, by being expeditious she may be put with them, before they have missed her, and they may be hived in the usual way.

In all cases, whether you set a new hive in place of the old one or not, whenever a swarm returns, if other hives stand near on each side, they are quite sure to receive a portion of the bees, probably a few hundreds, which are certain to be killed. To prevent this, cover them until the bees have gathered on their own hive. This is another argument in favor of plenty of room between hives.

Should no queen be discovered during the issue or return of the swarm, she should be sought for in the vicinity of the hive, and returned if found ; and the swarm will be likely to issue several days earlier than if obliged to wait for a young queen.

When the old queen is actually lost, and the bees have returned to wait for a young one, they are ready to leave one or two days sooner than regular second swarms. Whether a greater number of bees in the old hive, generating more animal heat, matures the chrysalis queen in less time than a stock thinned by casting a swarm, or some other cause operates, I can not say. I mention it because I have known it to occur frequently.

A swarm unaccompanied by a queen, is scattered more than usual when flying.

In most cases where the queens are unable to fly, they are old, and past the age of usefulness, and it is not of much consequence if they are lost. They would die soon, in any event.

FIRST SWARMS CHOOSE GOOD WEATHER.

First swarms are commonly more particular in regard to weather than after swarms. They have several days from which to choose, after the royal cells are ready, and before the queens are matured; and they usually select a fair one. But here again are exceptions.

SWARMING. 175

EXCEPTIONS.

I have known first swarms to issue in a wind that kept every branch of tree and bush in such agitation that it was impossible to find any upon which they could cluster. After a few fruitless attempts they gave it up, and came down on "terra firma." This occurred after several days of rainy weather. The next day being pleasant, many swarms issued, almost proving that the wind on the preceding day had kept a part of them back. I have also known them to issue in a shower that beat many of them to the ground before they could cluster. In these cases the shower was sudden, the sun shining almost at the moment it began to rain. During a long period of wet, cloudy weather they seem to become impatient, and come out quite unexpectedly—contrary to all rules.

AFTER-SWARMS.

After-swarms are all that issue after the first, called second, third, etc., for convenience. They differ in their conduct from the first swarms, as also do some first swarms when the old queen has been lost, and they are led out by a young queen.

THEIR SIZE.

Second swarms are usually half as large as the first, the third half as large as the second, etc., with some variations. I give general features, noticing only the exceptions that occur most frequently.

WHEN EXPECTED.

Whenever, in a prosperous season, the first swarm is not kept back by foul weather, the first of the young queens in the old colony is ready to emerge from the cell in seven or eight days. The second swarm may be expected in about two days thereafter.

PIPING OF THE QUEEN.

On the morning of that day, or the evening previous, by putting your ear close to the hive, and listening attentively a few minutes, you will hear a distinct piping noise like the word *peep*, uttered several times in succession, and followed by an interval of silence. Two or more may be heard at the same time; one will be shrill and fine, another hoarse, short and quick. The first is made by the queen that has left her cell, the other by one or more that have matured, but are kept in the cells by the workers, after they have made an opening for their exit. The difference in the sound is probably caused by their wings being cramped by the walls of the cells. They are so little disturbed by the removal of a comb, that the piping is continued while you are looking at the very comb upon which they happen to be. This piping is easily heard by any one not actually deaf, and there is not the least danger of its being mistaken for any humming; in fact, it is not to be mistaken for any thing, even when it is heard for the first time. These notes can probably never be heard, except when the hive contains a plurality of queens.

I never failed to hear it, previous to any after-swarm, whenever I listened. The time that the piping commences will be later than specified, in some colonies, if the weather is cool, or there are not many bees left; it may be twelve or fourteen days after the first swarm.*

Also, the swarm may not issue in two or three days after you hear the piping. The longer the swarm delays, the louder will be the piping. I have heard it distinctly twenty feet, by listening attentively when I knew one was thus engaged. By putting the ear against the hive, it may be heard even in the middle of the day, or at any time

---

* When first swarms issue before the queens in the old hive have advanced much, as they sometimes do, the second swarms issue from twelve to sixteen days afterward.

before the issuing of the swarm. The length of time during which it may be heard, seems also to be governed by the yield of honey; when that is abundant, it is common for them to issue the next day, but when it is somewhat scarce, they will very often delay three or four days. In such instances, third swarms seldom occur.

Piping for third swarms may usually be heard the evening after the second has left, though one day commonly intervenes between their issues.

VARIATION IN TIME OF SWARMS ISSUING.

Here my experience is at variance with many writers, who say there will be an interval of several days between second and third swarms. I do not remember of any interval of more than three days, but I have known many to issue in less time, several the next day, and a few on the same day with the second. I once had an instance where a swarm lost its queen, on its first sally, and returned to wait for the young ones; when they were ready an uncommon number of bees was present, and three swarms issued in three days! On the fourth another came out and returned; the fifth day it left, making four regular swarms in five days. On the eighth day the fifth swarm left. Although I had never had a fifth swarm before, I expected this from the fact that I heard the piping on the evening subsequent to the fourth swarm. The piping continued in this hive from the evening previous to the first swarm, till the last one had left.

Occasionally piping may be heard and no swarm issue. The bees seem to change their mind about swarming, and kill their queens, or allow the eldest one to destroy the others, as is evident from the fact that sometimes swarms are indicated, and none issue. When the piping continues over twenty-four hours, they seldom fail to swarm.

I have known in a few instances piping to commence,
8*

while, as I supposed, the old queen was yet present, and had not left the hive (on account of bad weather); but a swarm issued soon after. Also, I have observed instances of piping when I supposed the old queen lost, at a time when no swarm had been lead out; and the colony reared young ones to supply her place. This occurred in or near the swarming season, and one or two issues resulted. One case was three weeks in advance of the season, and the swarm was about half the usual size. When a swarm has been out and returned at the last of the swarming season, it is much more likely to re-issue, than if it depended for a leader on an old queen, that had not been out. Such will often issue later in the season than any others. A few have come out as early after the first swarm, as the fourth or fifth day, but all these are exceptions to the general rule.

HOW AFTER-SWARMS ISSUE.

When after-swarms start, the appearance about the entrance is altogether different from that when first-ones issue, unless there is an unusual number of bees. I have said that for a little time beforehand such were in an apparent tumult, etc. But after-swarms seldom give any such notice. One or more of the young queens may sometimes be seen to run out and back several times in a few minutes, in a perfect frenzy, and sometimes fly a short distance and return before the swarm will start. Even after the swarm is in motion she may return and enter the hive a moment. The workers seem more reluctant to leave, than in first swarms, where a mother instead of a sister is leader. No doubt she finds it necessary to exert herself to induce as many as possible to leave with her. A person watching the issue of a second swarm under these circumstances, for the first time, and seeing the queen leave first, would very likely *guess* that she did so in all swarms.

NUMBER OF QUEENS.

After-swarms sometimes have as many as six queens. The one containing several, is usually the last from the hive. When nearly all mature at once, and the workers keep them confined,—feeding them of course,—they become strong enough to fly, while in the cells. In the confusion of swarming, the prisoners are forgotten, and they come out and leave with the rest.

DO NOT ALWAYS CHOOSE GOOD WEATHER.

These after-swarms are not very particular about the weather; heavy winds, a few clouds, and sometimes a slight sprinkle will not always deter them. Neither are they very precise about the time of day. Italians will issue before six A. M. on warm mornings, and after five P. M., and the black bees are often nearly as much out of season. These things should be understood, because when after-swarms are expected, of which the piping will give warning, it behooves us to watch them in weather, and at times, when first ones would not venture to leave.

THEY GO FURTHER BEFORE ALIGHTING.

It is essential that some one sees them issue, else it is often difficult to find the cluster. They are apt to go further from the parent hive than others; sometimes fifty rods, and then often settle in two places, high and inconvenient, that distance apart. Let me not be misunderstood; I do not say they all do so, or even the majority, but a greater proportion of these swarms will do so than of the first.

If they cluster in two places, a queen may be in each, and they will remain, and when you have hived one part, you may think you have them all. If one cluster is without a queen, they will join the other, if near; but when distant, they will very likely return to the old hive soon, unless put with the others.

PROPRIETY OF RETURNING THEM.

Much has been said about returning all after-swarms to the old stock. The advantages will depend on the time of issuing, the yield of honey, etc. It would be unusual to have many after-swarms without a liberal yield of honey for the time being, but the continuation of the supply is uncertain. If honey continues plentiful, second and even third swarms, if early, may be hived, and prosper. The apiarian here needs judgment and experience.

It is always best, if possible, to have good strong families. When after-swarms are late, it is safest to return them, as the old colony will need them to replenish the hive, and prepare for winter. Also, it will be less infested with worms when well provided with bees, and there are more chances of obtaining box-honey. But the process of returning such, requires some patience and perseverance. I have said that there may be a dozen young queens in the old stock. Suppose that one or more leaves with the swarm, and you return the whole, there is nothing to prevent their leading out the swarm again the next day. Therefore it is policy to retain the queens. It is the least trouble to hive them in the usual way, and let them stand till the next morning. This will save you the perplexity of looking for more than one queen, if there should be more, for all but one will be destroyed by that time.

There is a chance also for the parent hive to decide that no more shall issue, and allow all but one to be slain there. When this is the case, and you find the one with the swarm, you will have no further trouble. They should be returned as soon as the next morning, otherwise they might not agree, even in the old home. To return them, and find the queen easily, get a sheet or a wide board a few feet long; let one end rest on the ground, the other near the entrance that they may enter the hive without flying; shake the swarm out on the lower end, and they will commence running up towards the hive; the first one that discovers it will call the others. If they do not perceive it, which sometimes happens, scatter some of them near it, and they will soon be marching in the right direction, when you should look for and secure the queen, if possible. Piping, a few hours later, will give notice, if they intend to issue again. It is evident, if these directions are followed, that they cannot issue many times before their stock of royalty will be exhausted; and when but one queen remains, the piping will cease, and trouble be at an end.

To prevent these after-swarms, some writers recommend turning the hive over, and cutting out all the royal cells but one. This I have found impracticable with most stocks. Some of the cells are too near the top to be seen, consequently this cannot always be depended upon.

It is somewhat difficult to give a rule for returning these swarms. If I should say return all that issue after June 20th, some seasons might be so late, that a second swarm issuing July 10th might fill the hive and winter well, while in others the first swarms in June might fail to get enough. Also, June 20th in the latitude of New York City, is as late as July 4th farther north.

In sections where Buckwheat is raised to any extent, late swarms do more towards filling their hives, than where that is not an important crop.

THE MOTH WORM TROUBLES SMALL COLONIES.

Should it be thought best to hive after-swarms, and risk the chances, they should receive a little extra attention, after the first week or two, in destroying the worms; a little timely care may prevent considerable injury. They are apt to construct more comb in proportion to the number of bees, than others; consequently, such combs cannot be properly covered and protected. The moth has an opportunity to deposit her eggs on them, and will sometimes entirely destroy them.

UNITING.

Whenever these swarms issue near enough together, it is best to unite them. I have said that second swarms were generally half as large as the first. By this rule, two second swarms or four third, or one second and two third would contain as many as one first swarm; if the first and second are of ordinary size, I think it advisable always to return the third. But in large apiaries, it is common for them to issue without any previous warning, just as a first swarm is leaving, and crowd themselves into their company, seeming to be as much at home, as if they were equally respectable.

MORE TROUBLE.

When two or more of these after-swarms are united, they are apt to be much more troublesome than others. The bees of each swarm are strangers to the queens belonging to the others. Bees usually make it a rule when coming in contact with a strange queen, while their own is present, to imprison her, as before described. So many of the bees observe this practice that every queen is soon surrounded. Directly some of the bees want their own queen, and cannot find her; forthwith consternation prevails throughout the hive. They run to and fro, fly out and return, set up the call for a moment, then perhaps return to some of the mother stocks; or if by chance there is a newly hived swarm in the yard, that behaves decently, they will join that and get up an excitement there, just because they are in trouble at home. When there is but one queen, and she is at liberty, she has not the sedate majesty of her mother, but seems often to be elated with her position. She will sometimes fly off and return, at others go back to the mother stock when the swarm will follow, and the experiment come to a very unsatisfactory termination. Perhaps those that behave so foolishly, have so recently entered society, that they do not know what course of conduct is becoming to them. Whenever they behave in this manner, it is well to confine the bees to the hive—giving them air—and keep them prisoners a day or two, until thoroughly sobered. Then if they are without a queen, give them one, or the means of rearing one.

RULE.

It may be accepted as a rule, that all after-swarms *must* be out by the eighteenth day after the first. I never found an exception, unless it may be considered as such when a swarm leaves, seven or eight weeks after the first. But these I consider rather in the light of first swarms, as they issue under similar circumstances, leaving the combs in the old hive filled with brood, queen cells finished, etc. A hive may cast swarms in June, and a buckwheat-swarm in August, on the same principle.

Therefore, bee-keepers having but few hives, will find it useless to watch their bees, when the last of the first swarms came out sixteen or eighteen days before. Much trouble may be thus saved by a little knowledge of facts. During my early days in bee-keeping, I was anxious for the greatest possible increase of stocks. I had some that had cast a first swarm, and soon after, clustered out again. I watched them vainly for weeks and months, expecting another swarm. But, had I understood the "modus operandi" as the reader may now understand it, my anxiety as well as watching, would have been at an end in a fortnight. As it was, it lasted two months. I found no one to give me any light on the subject, or even tell me when the swarming season was over, and I came very near watching all summer!

ONE QUEEN DESTROYS OTHERS.

When it is decided in family council, that no more swarms are to issue, all but one of the queens are destroyed. It is probable that the oldest and strongest dispatches the others, while in the cells, or allows them to issue, and take a fair fight.

When rearing Italian queens in the small boxes, it is usual to have half a dozen queen-cells on a very small piece of comb. To save these from destruction, all but one must be cut out before any hatch. If the brood given them is just the right age—about four days old—a queen will hatch in ten days, and if the others are not removed, the first one that hatches, makes it her business to destroy the rest. I have often caught them when just out of their own cell, at work at the others. The younger sisters in helpless confinement are slaughtered without mercy. An opening is bitten into the royal cell, and the fatal sting inflicted in the abdomen of the defenceless queen.

If quick and spiteful movements are any indication of hatred, it is here very plainly manifested. The bees enlarge the opening and drag out the dead queens.

It is probable that all swarming hives manage in this way when it is decided to send out no more swarms, as we find numbers of dead queens about the entrance just at this time; and this may generally be taken as evidence that swarming is over in such hive for the season. Should the stock send out but one swarm, the dead queens may be found about the time, or a little before, you would listen for the piping.

Whenever hives containing swarms are full, or nearly so, boxes should be put on without delay, unless the honey season is so nearly over that it is unnecessary.

CHAPTER XI.

ARTIFICIAL SWARMS.

Artificial swarms are those which are made by driving or dividing. The utility of such swarms will depend greatly on the circumstances of the bee-keeper; the time that he has to attend to regular swarms, and his general knowledge of the subject. There are advantages as well as disadvantages. There is not much difference between the labor of making artificial swarms, and of hiving regular issues. If I were sure of but one issue from a hive, and could always attend to the hiving without particular inconvenience, I would prefer natural swarms. But when we depend on these, and perhaps feel particularly anxious for them to issue, some will pertinaciously adhere to the old stock through the whole swarming season. When we have but few hives, and are particularly anxious to increase the number, this indifference to our wishes is very annoying. The other extreme—over swarming—is often still more vexatious.

PERPLEXITIES.

There are likewise some perplexities with artificial swarms. We do not always take out the requisite number, or we get too great a proportion of old or young bees, and when they are thus improperly divided, they do not always work well at first. One writer says, "artificial swarms, so called, I do not approve of at all, they do not work like the others." I cannot imagine why he should have failed, unless there was a lack of the requisite number of workers in all the departments, such as nurses, wax-workers and gatherers. Whether there is an organic distinction in the bees that fill these stations, or only temporary details for the purpose, I shall not express an opinion.

I know that very young bees act as nurses, but I presume, that as they grow older, they will, if they have an ordinary share of energy, go abroad and collect honey and pollen.

WORK WELL.

Artificial swarms do work *just as well as natural ones*, as a general rule. In fact I *never* had one, that I thought was less industrious, because of the manner in which it was made.

If you wish to be *sure* of an annual increase, it will be necessary to take the matter in part in your own hand, and make each hive spare a swarm that is in condition to do so. When this is decided upon prompt action is necessary.

DO IT IN SEASON.

It will not do to "wait and see if they don't swarm," and then do it, and then if they do not fill the hive and store as much surplus as a natural swarm hived four weeks sooner, attribute it to the manner of making the colony. Do it in season, or not at all. Also, it is important that a swarm is not taken at any time, unless the colony is abundantly able to spare it. The ability to decide this point requires much observation and experience. It should always be done when there is plenty of honey, unless you expect to feed, and it is usually safer to perform the operation during the swarming season. Without these conditions it is much better to postpone artificial increase till another year.

MY FIRST EXPERIENCE.

My first experience in making artificial swarms, and in raising queens was not very encouraging. But by complying a little more with the *natural* requirements of the bee, I have since succeeded satisfactorily. It is stated by nearly every writer, that whenever a colony of bees possessing eggs or young larvæ is deprived of its queen, they will not fail to rear another. This may be taken as a rule, but there are exceptions. The first experiments that I made in this line, came very near proving to me that the exceptions formed the rule. Very soon after I began to keep bees, when I had but few stocks, and was anxious to increase the number, I was perplexed with the failure of some hives to swarm, notwithstanding they were well supplied with bees, and exhibited the usual indication, such as clustering out, etc. Others, apparently not so well supplied with bees, threw off swarms. Taking the assertions of these authors for facts, I reasoned. thus: In all probability there is plenty of eggs and brood in each of those stocks. Why not drive out a portion of the bees with the old queen, and leave about as many as if a swarm had issued? Those left will then raise a queen, and continue the old stock, and I shall double the number. On examination, I found eggs and larvæ, and accordingly divided them. Of course, all *must be right*. Now, thought I, my stocks can be doubled, at least annually. If they do not swarm, I can drive them.

My swarms prospered, the old stock seemed industrious, bringing in pollen in abundance, which, at *that* time, was conclusive evidence that they had a queen, or soon would have. I continued to watch them with much interest, but somehow, after a few weeks, there did not seem to be as many bees in the old hives; a few days later, I was quite positive of it. I examined the combs and behold! There was not a cell containing a young bee of any age, nor even an egg in any of these old stocks. My visions of future increase by this means, speedily disappeared about this time.

My new swarms, it is true, were in condition for winter, although not full; but the old ones were not, and nothing was gained. I had some honey and a great deal of beebread and old black comb. Had I let them alone and put on boxes, I should probably have obtained twenty-five or thirty pounds of pure honey from each; besides, the old stocks, even with old comb, would have been better supplied with both honey and bees, and altogether much better stocks for wintering. Here was an important loss, arising simply from ignorance.

I looked the bees over carefully, and ascertained to a certainty that none of them had a queen. The few bees left, I smothered in the fall. I then knew of no better way. I had been told that the barbarous use of "fire and brimstone" was part of the "luck"—that a more benevolent system would cause them "to run out," etc. I cannot, to this day, account for my want of success. Since then, I have succeeded nineteen times in twenty, under circumstances, apparently precisely similar.

The swarming season is certainly the best time, as then most of the stocks are constructing these cells, preparatory to swarming, and there can hardly be a failure with the method recommended. But I shall advise furnishing the old stock with a queen before they can raise one, either by giving them a cell ready to hatch, or a laying queen. It is very plain that a queen from a finished cell must be ready to deposit eggs several days sooner, than one which is raised in the hive, after the necessity for one exists.

It is also clear, if we have a dozen queens depositing eggs by June 10th, that our bees are increasing faster on the whole, than if but half that number are engaged in it for a month later. There is yet another advantage. The sooner a young queen can take the place of the old one in maternal duties the less time will be lost in breeding, the more bees there will be to defend the combs from the moth, and the sooner the guarantee for surplus honey.

HOW TO MAKE ARTIFICIAL SWARMS.

When you are all ready, take a stock that can spare a swarm; if bees are on the outside, raise the hive on wedges, sprinkle them with a little water to drive them in, and disturb them gently with a stick. Now smoke and invert it, setting an empty hive over. If the two hives are of one size, and have been made by a workman, there will be no chance for the bees to escape, except through the holes in the side, these you will stop. With a light hammer or stick strike the hive a few times lightly, and let it remain five minutes. This is very essential, as it allows the bees to fill themselves with honey. All regular swarms go forth so laden. A supply is necessary when bad weather soon follows. The amount of honey carried out of a stock by a good swarm, together with the weight of the bees, (which is not much,) varies from five to eight pounds.

When the bees have filled their sacks, proceed to drive them into the upper hive, by striking the lower one rapidly from five to ten minutes. A loud humming will mark their first movement. When you think half or two-thirds are out, raise the hive and make an examination. They are not at all disposed to sting in this stage of the proceeding, even when they escape outside. If full of honey, they are seldom provoked to resentment. The only care necessary, is not to crush too many between the edges of the hives. The loud buzzing is no sign of anger. If your swarm is not large enough, continue to drive until it is. When done, the new hive should be set on the stand of the old one. A few minutes will decide whether you have the queen with the swarm, as they will remain quiet if she is present; but if she is not, they will be uneasy and run about, when it will be necessary to drive again.

MANNER OF PLACING THE STANDS.

If both hives are one color, set the old one two feet in front, but if of different colors. a little farther. This is for the box hives.

If there is plenty of room, the two hives can be set each side of the old stand, about one foot from it. As this should be performed in the middle of the day, you should set an empty hive, resembling the old one, on the stand, to catch the bees returning from work during the operation. After the hives are in their proper places, the bees in this temporary hive should be shaken out before them. The hive that receives the most of the returning bees should be set a little farther from the old stand, and the other a little nearer, until they enter in equal numbers, or this method may be adopted. When the hive to be divided was set so far from all others in the spring, that returning bees will not enter neighboring hives, you may leave the old hive on its own stand, and place a new one for the swarm, three feet from it, on one side. When the bees are divided, put the old hive in the cellar for a few days, until all the bees belonging to the new hive have become habituated to it, when the old hive may be returned to its own stand, and all the bees that are in it, will of course adhere.

INTRODUCING QUEEN-CELL.

Before you turn over the old stock, look as far as possible among the combs, for queen-cells; if any contain larvæ, you may leave them to rear a queen; but if otherwise, wait twenty-four hours, and then go to a stock that has cast a swarm, or to one of the little queen-boxes and obtain a finished royal cell, and introduce it. When there are young queens in the cells at the time of driving, after-swarms may issue. Should a queen-cell be introduced immediately, it is more liable to be destroyed than after an interval of twenty-four hours, and even then, it is not *always* safe. After it has had time to hatch, which is not far from eight days after being sealed, cut it out, and examine it: if the lower end is open, it indicates that a perfect queen has left it, and all is right; but if it is mutilated or open at the side, it is probable that the queen was destroyed before maturity, in which case it will be necessary to give them another cell.

OPERATION WITH MOVEABLE COMBS—EASY.

It is much more pleasant to operate with the movable combs than with the common hive. To divide, you have only to get an empty hive of the same size as the one you wish to divide. Place your stands one on each side of the old one, with the hives upon them. Begin two or three frames from the one you design taking out first, by moving them away from it a little, then take hold of each end and raise it carefully, without striking the ends or other frames. When the ends of the frame rest on the rabbeting of the hive, the bees seal them fast with propolis. A small chisel or bit of iron will be necessary with which to pry them loose. Loosen all, before lifting out any. The bees resent a slight jar during this operation more than the removal of the frames, and will need a little more smoke to quiet them. Take out just half the combs with the bees attached, and put them into the empty hive. If you have no empty combs, fill out each hive with empty frames. See that the bees enter them equally. This may be regulated in the same manner as with the box-hive.

ONE DIVISION WILL MAKE DRONE COMB.

The division, containing the queen, will, if they are obtaining honey, commence worker-combs; the other, will commence drone combs, and at the same time will be likely to make queen-cells on the old combs. The absence of a queen may be ascertained by these indications. Should the one with the queen contain queen-cells, started before the division, there might not be any combs made, and a swarm would be quite sure to issue as soon as any cells were finished. I once had three swarms from each half of a divided hive.

TOO MANY DRONE COMBS FOR PROFIT.

It would be well to ascertain as soon as the next day which half has the queen, and take off any cells begun in her apartment. In the other hive, before they have hatched a queen, they will make altogether too many drone combs for profit, in a stock-hive.

The honey now stored in these combs is in bad shape, and unless wanted for home consumption, some measures should be taken to have it in suitable shape for market. To cut them out, and transfer to the surplus boxes is attended with much trouble and some waste. I prefer to have it made directly in the boxes by the bees. But as they will not now go to the top of the hive to fill boxes, I put them inside in this way.

HONEY MADE IN BOXES IN THE HIVE.

Put a cross-piece some two inches wide in the centre of the frame and another at the bottom, thus.—Set on these, boxes of the usual size, with the glass removed on the side, next the comb and bees. The guide-combs should be put in parallel with the other combs, else they might be worked fast to them.

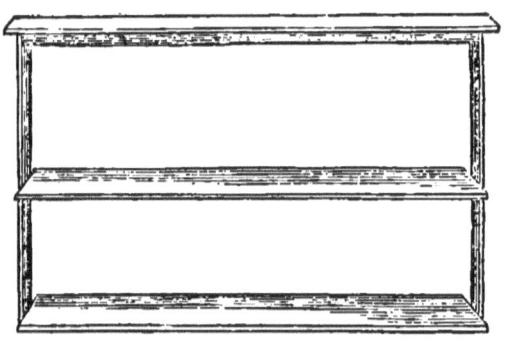

Fig. 24.—FRAME TO HOLD BOXES INSIDE THE HIVE.

If your hive is of the right size, you will have no difficulty, but if too shallow, as most movable comb hives are, it may be necessary to make surplus boxes for the purpose.

My hives happened to be of the right size—when made without reference to this process. They should be twelve inches wide, admitting frames 11x18; box about six inches by five deep. The boxes should be wedged firmly in the frame. These, with the frame, will, when properly adjusted, fill one-half the hive, and the four frames with combs, the other half. The open side of the box allows the bees to work with the same facility as in making additional combs in the hive. There is no danger of finding brood in them; and the labor expended is in a profitable direction.

BOXES TRANSFERRED AND FINISHED ON ANOTHER HIVE.

When the young queen is mature, and commences laying, they should be removed. If not full, and the colony is strong, put them on the top, and the bees will remain and finish them. But if the colony is weak, they will leave to work below, when they should be transferred to some other hive to be finished.

TIME FOR QUEEN TO LAY EGGS.

The young queen will make her appearance on the twelfth day from the time the division is made; in eight days more she will commence laying. This is the rule, but exceptions are frequent. If you wish to use the surplus queen-cells that will probably be made, for other divisions, or any purpose, they should be taken out by the tenth day. The queen first hatched sometimes destroys them all by that time. When all goes well, the queen should be laying in three weeks. Eggs in the cells at that time indicate her presence. When a queen-cell can be given to the half, destitute of a queen, the young queen is ready to lay several days earlier. When a laying queen can be introduced, there is an average gain of from ten to fourteen days.

9

I will give another method of dividing with movable combs, which some may prefer to adopt.

Remove one frame containing brood comb, with the bees that happen to be on it, together with the queen, to an empty hive and place it on the old stand, setting the old stock four feet to one side. Fill out the new hive with empty frames. Most of the bees will come to this stand, but enough will usually remain in the old hive, in addition to the brood that hatches out, to keep it in a prosperous condition. The frames should be moved together, and the vacancy made by the removal, filled by an empty frame at the outside. The hive being full of combs, they will not construct new. They may be furnished with a queen-cell as before directed. The objection to this method is their occasional disposition to swarm out.

When the trouble is no great obstacle, and it is desired to increase the stocks to the utmost, without regard to surplus honey, perhaps more good swarms may be made, by having the breeding hive rather large, and keeping the old queen at home, continually supplying the combs with eggs. If she is usually prolific, a swarm of more than 20,000 bees may be taken out every three or four weeks. Take out the combs, shake off and get into an empty hive all the bees proper to spare, and return the old queen. Give them a mature queen in a wire-cloth cage, as described in Chap. XXIV. Confine the bees a few hours, and remove at least a mile. The queen may be liberated after three or four days. Her presence will generally prevent much drone-comb from being built. If any comb of the old hive becomes filled with honey, it should at once be removed and replaced with an empty one, as it will occupy room that should be used for brood. The advantage of this plan is, that all the combs of the parent hive, are maturing bees throughout the season, and there is no loss of time, as for some weeks in swarming hives.

CHAPTER XII.

LOSS OF QUEENS.

If all my readers were keeping bees in the movable-comb hive, I should have but little to say in this Chapter, but as the box-hive will yet be used, it will be necessary to give directions accordingly.

WHEN LOST BY SWARMS.

Swarms that lose their queens in the first few hours after being hived, generally return to the parent stock; sometimes unite with some other. If much time has elapsed before the loss, they remain, unless standing on a bench with another. On a separate stand, they continue their labor, but a large swarm diminishes rapidly, and seldom fills an ordinary hive.

DRONE COMB.

A singular phenomenon attends a swarm that is constructing combs, without a queen. I have never seen it noticed by any one, and it may not always be the case, but I have so found it in every instance that has come under my notice. That is, four-fifths of the comb is composed of drone-cells; why they thus construct them is a subject for speculation, from which I will endeavor in this instance to refrain.*

SPECULATION.

It has been suggested as a profitable speculation "to hive a large swarm without a queen, and give them a piece of brood-comb containing eggs, to rear one, and as soon

---

* The above was written 12 years ago. About the same time, Mr. Langstroth noticed the same fact, in his work, and it is now pretty generally known, but not yet explained.

as it is matured, deprive them of it, giving them another piece of comb, and continue it throughout the summer, putting on boxes for surplus honey. There being no brood to consume any honey, no time will be lost, or taken to nurse them, and as a consequence, they will be enabled to store large quantities of surplus honey."

This appears very plausible, and to an inexperienced person somewhat conclusive. If success depended on some animal whose lease of life was a little longer, it would answer better to calculate in this way. But as a bee (the queen excepted) seldom sees the anniversary of its birthday, and most of them perish in a few weeks or months, it is bad economy. It will be found that the largest amounts of surplus honey are obtained from the prolific stocks. Therefore, it is all important that every swarm and stock has a queen to repair this constant loss of bees.

DISPUTED QUESTION.

We now approach a disputed point in natural history, relative to the queen leaving at any time, except when leading out a swarm. Most writers say that the young queen leaves the hive, and meets the drone on the wing. Others *positively* deny this, having watched a whole summer without seeing her leave. Consequently, they have arrived at the very plausible, and apparently consistent conclusion, that nature never intended it to be so, since it must happen at a time when the existence of the whole family depends entirely on the life of the queen. The stock at such times contains no eggs or larvæ, from which to rear another, if she should be lost. "The chances at such times of being devoured by birds, blown away by the winds, and other casualties, are too many, and it is not probable the Creator would have so arranged it." But facts are stubborn things; they will not yield one jot to favor the most "finely spun theory;" they are most provokingly obstinate, many times. When one takes a survey of animated nature, and finds that male and female are usually about equal in number, he is ready, and often does conclude that a single bee among thousands can not be the only one capable of reproduction or depositing eggs. The idea, to him, is preposterous! And yet, only a little observation will upset this apparently consistent reasoning. So it seemed to be very inconsistent that the young queens should leave the hive, but I was compelled, though reluctantly, to admit it. That this purpose is, to meet the drones, cannot at present be contradicted. Also, that when the queen is once impregnated, it is operative for life. She is never detected coming out again for that purpose.

A MULTITUDE OF DRONES NEEDED.

What then is the use of the ten thousand drones that never fulfill this important duty? It seems indeed like a useless expenditure of labor and honey for each stock to rear thousands, when perhaps, but one, sometimes not any, of the whole number is of any use. If the risk is great in the queen's leaving, we find it admirably arranged that it shall not be too frequent.

Instinct teaches the bees to make the matters left to them, as *sure* as possible. When they want one queen, they raise half a dozen. If one drone, or only half a dozen were reared, the chances of the queen meeting one in the air, would be very much reduced. But when a thousand are in the air, the chances are a thousand times multiplied. If a stock casts a swarm, a young queen must be impregnated and return safely, or the stock is lost. Every time she leaves, there is one chance in about twelve of her not returning. If the number of drones were any less than it is, she would have to repeat her excursions, till successful. As it is, some have to leave several times. The chances and consequences of loss are so great, that on the whole, it is no doubt better to rear a thousand unnecessarily, than to lack one in time of need. Therefore, let us be content with the present arrangement, inasmuch as we could not better it, and probably, should we try, would " so fix the thing that it would not go at all."

But what is the use of drones in hives that do not swarm, and do not intend to do so, as in very large ones, or those situated in a large room? In such circumstances they seldom produce swarms, yet as regularly as the return of the summer, a brood of drones appears. What are they for? Suppose the old queen in such hive dies, leaving eggs or young larvæ, and a young queen is reared to supply her place. How is she to be impregnated without the drones? Perhaps they are taught that whenever they *can* afford it they should have some on hand to be ready for an emergency. I have already said that when bees are numerous, and honey abundant, they never fail to provide them. A crippled queen hived with a swarm, or even in an old stock, is generally replaced by one that is perfect, within a month or two.

WHEN THE LOSS OCCURS.

Whenever I have witnessed this excursion of the queen, it has taken place a little after the middle of the day, when the drones were out in the greatest numbers. At such times, there is rather more than usual commotion among the workers. I have watched their return—their absence varying from three minutes to half an hour—and have seen them hover around their own hive, apparently in doubt whether they belonged in that, or the next; in a few instances they have actually settled on the neighboring hive, and would have perished there, but for my assistance. Thus we see that queens are lost on these occasions, from some cause; and part, perhaps most of them, by entering the wrong hive; if so, it is another good reason for not setting hives too closely together. The hives are very often nearly alike in color and appearance, and the queen coming out for the first time in her life, is doubtless confused by this similarity.

The average number of such losses in a season, varies: One year, the average was one in nine, another, one in thirteen, and another, one in twenty. The time after the issuing of the first swarm varies from ten to twenty days. The inexperienced reader should not forget that these accidents happen in the old stocks which have cast swarms, the old queen having left with the first. Also, all after-swarms are liable to the same loss.

I would suggest that these have abundant room given between the hives; if it be necessary to pack any closely, let it be the first swarms, where, the queen, being old, has no occasion to leave. Having never seen this matter fully discussed, I wish to be somewhat particular, and think I shall be able to direct the careful apiarian how to save a few stocks and swarms annually. Several years ago I wrote an article on this subject, for an agricultural paper. A subscriber told me a year afterwards, that he saved two stocks the next summer by the information; they were worth, at least five dollars each, enough to pay for his paper for years to come.

TIME OF LEAVING VARIES.

When a stock casts but one swarm, the queen, having destroyed all competitors who would interfere with her movements, will leave in about fourteen days, if the weather is fair; but should an after-swarm leave, the oldest of the young queens will probably go with that. Of course, then, it must be later before the queen remaining in the old hive is ready; it may be twenty days, or even more. The queens with after-swarms will leave from one to six days after being hived. It always will occur when no eggs or larvæ exist, and no means left to repair the loss. A loss it is, and a serious one; the bees are in as much trouble as their owner, and quite likely, more, as they seem to understand the consequences, and he is perhaps ignorant; should he now for the first time learn the nature of it, he will at the same time understand the remedy.

INDICATIONS OF LOSS.

The next morning after a loss of this kind has occurred, and occasionally at evening, the bees may be seen running to and fro in the greatest consternation on the outside. Some will fly off a short distance, and return; one will run to another, and then to another, still in hopes, no doubt, of finding their lost queen. A hive, close by on the same bench will probably receive a portion, and will seldom resist an accession under such circumstances. All this will be going on while other hives are quiet. Towards the middle of the day, the confusion will be less marked; but the next morning it will be exhibited again, though not so plainly, and will cease after the third, when they become apparently reconciled to their fate. They will continue their labors as usual, bringing in pollen and honey. Here I am obliged to differ with writers, who tell us that all labor will now cease. I hope the reader will not be deceived by supposing that the collection of pollen is an *infallible* indication of the presence of a queen. I can assure him it is not always the case.

RESULT.

The number of bees will gradually decrease, and they will be gone by the early part of winter, leaving a good supply of honey, and an extra quantity of bee-bread, as before mentioned, because there has been no brood to consume it. This is the case where a large family is left at the time of the loss. When but few bees are left, it is very different; the combs are unprotected by a covering of bees, the moth deposits her eggs on them, and the worms soon finish up the whole. The bees from the other stocks will generally first remove the honey. Hundreds of bee-keepers lose some of their stocks in this way, and can assign no reasonable cause. "Why," say they, "there wasn't twenty bees in the hive; it was all full of honey," or worms, as the case may be. "Only a short time before, it was full of bees; I got three good swarms from it, and it always has been first-rate, but all at once the bees were gone. I don't understand it!"

AGE OF BEES.

Such bee-keepers do not understand how rapidly a family of bees diminishes, when there is no queen to counterbalance with young, this regular decrease. I doubt whether the largest and best family could possibly be made to exist more than six months, without a queen for their renewal, except perhaps during the winter.

DUTY.

As this tumult can be seen but a few days at most, it is well, yes, *necessary* to make it a duty to glance at the hive *every morning*, at this period after swarming; a glance is sufficient to discover the fact. Remember to reckon from the date of the first issue; this occurs when the first royal cells are sealed over, and is the best criterion by which to judge when the queen will leave.

REMEDY.

When a loss is discovered, first ascertain if there is any after-swarm to be expected from another stock, by listening for the piping; if so, wait till it issues, and obtain a queen from that, for your stock. Even if there is but one, take it, and let the bees return; they will probably come out again the next day, if they do not, it is very often no great loss. Should no such swarm be indicated, go to a stock that has cast a first swarm within a week, smoke it, and turn it over, as before directed, find a royal cell, and cut it out, being careful not to injure it. This must now be secured in the other hive, in such a position that the lower end will be free from any obstacle, which will interfere with the egress of the queen. It will make but little difference whether at the top or bottom, if it be secure from falling, and can be kept warm by the bees. I generally introduce it through a hole in the top, taking care to find one that will allow the cell to pass down between two combs. Being largest at the upper end, the combs each side will sustain it, and leave the lower end free. In a few hours the bees will secure it permanently to the combs with wax. This operation cannot be performed in a chamber hive, as it is impossible to see the arrangement of the combs through the holes. To put it in at the bottom is more trouble. The difficulty is, to fasten it, and prevent its resting on the end. It may be done as follows. Take a piece of old tough comb an inch square. Make a hole through the centre large enough to receive the cell, turn up the hive, and spread two combs far enough apart to receive the piece between them, which arrangement will secure it from falling.

The bees will become quiet, soon after such cell is introduced. It will hatch in a few days, and they will have a queen as perfect as if it had been one of their own rearing. This queen, of course, will be under the necessity of leaving the hive, and will be just as liable to be lost as others, but no more so, and must be watched as carefully.

It is unnecessary to look for a cell in a stock that has cast its first swarm more than a week before, as they are generally destroyed in that time, sometimes, in less, unless they intend to send out an after-swarm. When artificial, or laying-queens are kept on hand, no one need be told to introduce such a one at once.

MARK DATE OF SWARM.

Should the apiary contain so many stocks that it is difficult to remember the date of each swarm, it is a good plan to mark it on one side or corner of the hive, as the swarms issue. It will thus be easy to tell where to look for a cell.

OTHER REMEDIES.

It will sometimes happen that a queen is lost at the extreme end of the swarming season, when no stock contains such cells, and no queen is at hand. In such a case, it is often economy to take a queen from the most inferior stock on hand, and sacrifice it to save the queenless one. If no poor colony is at hand, drive the bees out of one of the best, secure the queen, and return the bees. They will raise another, and the damage will be less than to lose the queenless stock. The strong one will recover, but the other needs a queen, at once, and cannot afford to take time to raise one. Therefore I would recommend introducing a mature queen whenever it is practicable. When all the brood in the combs is hatched, and the bees are obliged to commence with an egg to raise a queen, there can be no young bees added to the colony short of six weeks, by which time, most such would be beyond recovery. Sometimes after all our efforts, a few stocks will remain destitute of queens. These, if they escape the worms, will generally store honey enough, in this section, to winter a good colony. This must be introduced of course, from another hive containing a queen, but this belongs to Fall Management.

INDICATIONS OF LOSS IN EARLY SPRING.

Occasionally, a queen is lost, at other than the swarming season, averaging about one in forty cases. It is most frequent in spring, at least, it is generally discovered then. The queen may die in winter, and the bees give no indications of it until they come out in spring. Occasionally, they may all desert the hive, and join another. If we expect to ascertain when a queen is lost at this season, we must notice them just before dark on the first warm days, because the mornings are apt to be too cool for any bees to be outside; any unusual stir or commotion, similar to what has been described, indicates the loss. This is the most difficult time of the year to provide the remedy, unless there should happen to be some very poor stock containing a queen, that we might lose any way, which it would be judicious to sacrifice to save the other, especially if the latter contains all the requisites of a good stock, except a queen. As soon as drones appear, it would do to take a queen from a strong stock, as just mentioned. In such a case, the movable comb hive is an advantage. Combs from some full hive containing considerable brood, occasionally introduced into a queenless one, will be a great help to the colony, and keep it in a thriving condition, until a queen can be procured. They will probably raise one immediately on receiving the brood, but if it be too long before there are drones, she will prove a drone queen, and must be destroyed, and another substituted. If empty combs can be supplied to the hive from which such brood combs are taken, scarcely any difference will be observed in their prosperity.

Thus far in my experience with the Italians, I have observed that they seldom lose their queens.

CHAPTER XIII.

PRUNING.

This chapter, like some others, would be useless, were it not that the box hive is yet much in use.

SELDOM NECESSARY.

The apiarian whose main object is profit, will find that pruning is desirable much less often than the patent vender recommends, and in sections where foul brood exists, it is very seldom necessary. Yet many will like to know how it should be done.

THE TIME.

The time at which it should be performed, is of some importance. The month of March has been recommended by many, others prefer April, August or September.

Here, as usual, I shall differ from them all, preferring still another period, for which I offer my reasons, supposing of course, that the reader is conscious of a freeman's privilege of choosing whatever time or method he thinks proper, in this, as in other matters There is but one period from February till October, when prosperous stocks are free from young brood in the combs. If combs are taken out when thus occupied, there must be a loss of all the young bees they contain. The old queen leaves with the first swarm; all the eggs she leaves in the worker-cells will be matured in about twenty-one days; hence *this* is evidently the best time to prune the old combs with the least waste. A few unhatched drones will be found in the cells, but they are of no account. Also, a few very young larvæ and some eggs may sometimes be found, the product of the young queens; these must be wasted, but as the bees have expended no labor upon them, it is better to sacrifice them than the greater number left by her mother, which have consumed their portion of food, and have been sealed up by the bees. Should this operation be postponed more than three weeks, the young queen will so fill the combs again, as to involve a serious loss. Therefore, I wish to urge the necessity of attention to this point at the proper season. If you think it unimportant to mark the date of your first swarms, for the purposes mentioned elsewhere, it will be found very convenient here, for those that need pruning.

It is recommended by some, to take out only a part, say one-third or half of the combs in a season, thereby taking two or three years to renew them. This is advisable only when the family is very small. As this space made by pruning cannot be filled without wax and labor, our surplus honey will be proportionate to its extent. Suppose we take out half the old combs, and get half a yield of surplus honey this year, and do the same next, or complete the operation, and have none this year, and a full yield next. What is the difference? There is none in regard to honey, but some in trouble, and it is in favor of performing the whole operation at once. Besides the advantage of saving a large brood by pruning at this season, such hives will usually refill before fall, and are much better for wintering, than if done later in the season. In the latter case, much brood will be wasted, and a large space will be unoccupied with combs during the winter. But few combs can then be made, and those few must be at the expense of their winter stores, unless we resort to feeding. These objections apply with still greater force to pruning in March or April. The loss of brood is of much more consequence then, than in mid-summer, or even later; and a space to be filled with combs is a serious disadvantage. It is important that the bees should devote their whole attention to rearing brood, and be ready to cast their swarms as early as possible. One *early* swarm is worth two late ones.

DIFFICULTY IN DRIVING IN COOL WEATHER.

Further, it will be found much more difficult to drive the bees out of a hive in the cool weather of March or April, than in summer, as they seem unwilling to leave their warm quarters and go into a cold hive. The first thing necessary, is to get rid of the bees, and the operation of pruning is performed much quicker when they are driven out in the outset. If there are not bees enough in the hive to interfere, it will not pay to prune the combs.

BEST TIME.

The best time to begin is just long enough before to complete it by dark. First, blow some smoke under the hive, turn it over, and set on it an empty hive the same size. Stop all crevices, and rap on the lower hive a few times, with a light hammer or stick. The bees becoming alarmed, will set up a loud buzzing, and most of them will commence filling themselves with honey. Proceed with the drumming, and when they have taken all the honey they can carry, they will readily ascend into the upper hive. The loud buzzing is not so much a sign of anger as of fear. In five or ten minutes, one edge of the upper hive may be raised to inspect progress. When most are up, set the upper one on the old stand, get another empty one, and drive out more, shake these down in front of the others, and they will immediately enter. If there are only a few left scattered here and there on the combs, they may be disregarded. By this time, none of them are disposed to sting, unless they have Italian blood in them.

Should it be desired to drive the bees out permanently, for reason of diseased brood, or other causes, you have only to continue the process until all are out. They will go to work as a new swarm. The reason that complaints are made of such swarms not doing well generally, seems to be in allowing the colony to decrease too much before driving, thus leaving too *few* bees to accomplish anything.

The tools for pruning are very simple. The broad one is readily made by any blacksmith, from a piece of an old scythe, about eighteen inches long, by simply taking off the back, and forming a shank for the handle at the heel. The end should be ground all on one side, and square across like a carpenter's chisel. This is for cutting the

Fig. 25.—TOOLS FOR PRUNING.

combs at the sides of the hive; the bevel will keep it close the whole length, when you wish to remove the whole of a comb. Being square instead of pointed or rounded, no difficulty will be found in guiding it, and being very thin, no combs will be broken or crushed. The other tool is for cutting off combs across the top, middle, or any place where it is desired to cut horizontally. It is merely a rod of steel three-eighths of an inch in diameter, about two feet long, with a thin blade at a right angle, one and a half inches long, one-fourth inch wide, both edges sharp, upper side bevelled, bottom flat, etc. These are convenient for many purposes besides pruning, and the cost cannot be compared to the advantages. Now, with these tools, proceed to remove the brood combs from the centre of the hive to be pruned. The combs near the top and outside are used but little for breeding, and are generally filled with honey; these should be left as a good start towards refilling. Reverse the hives, putting the one containing the bees under the other; by the next morning all are in their own hive. Put it on the stand, and the work is done without any extra expense for a patent, and the bees are much better off for the honey left, which must be taken away, with all patent plans that I have seen, except the movable combs. This is worth much more to the bees than to the owner, as it often contains cocoons and bee-bread, and they will repay with pure comb and honey.

LITTLE RISK OF STING.

The general objection to this mode of renewing combs, is the fear of being stung. There is, however, but little danger, not as much as in walking among the hives in a warm day. Begin properly, use smoke, work carefully, without pinching them, and you will generally escape unhurt. With the movable combs, it is only necessary to take out a comb, shake off the bees, cut out what comb you wish, and return it to the hive.

FREQUENT PRUNING NOT RECOMMENDED.

In giving these directions, I do not wish it to be forgotten that I disapprove of frequent pruning. Combs once used for breeding, can be used for no other purpose as well, as they are undesirable for storing honey. The time of the bees can be much more profitably employed than in building new brood-combs every year. Combs can be used for ten years without detriment to the bees. The idea that bees will become dwarfed by being raised in cells long used for such a purpose, is seldom entertained by practical bee-keepers. I have long believed it impossible for the most interested advocate of renewal, to detect any diminution in the size of such bees.

CHAPTER XIV.

DISEASED BROOD.

WHAT IS IT?

I find since writing the original chapter on this subject, that bee-keepers are much more familiar with it than I supposed. Mr. Langstroth in his work gave a fair description, terming it "foul brood," but knew nothing of it from his own experience. Mr. S. Wagner, of York, Pa., has found much on the subject in the German Bee Journals. He has kindly sent me translations of some articles contributed by Dzierzon, the great German apiarian. I find but little that is not identical with the experience concerning it in this country. In its first appearance in Dzierzon's apiary it was much more disastrous than I ever knew it to be, and it came near sweeping away his whole apiary. Afterwards, when better acquainted with it, he was more successful in his management.

ITALIANS LESS AFFECTED.

It would be interesting to learn how he succeeds, since he has introduced the Italians. Since their introduction into my apiaries, the number affected with this disease has diminished five-sixths. Whether this is to be attributed to them I cannot say, but I am inclined to give them some credit for it.

This disease is probably of recent origin in this country. Mr. Weeks said in a newspaper article some years ago, "since the potato rot commenced, I have lost one-fourth of my bees annually by this disease," and adds his fears, "that this race of insects will become extinct from this cause, if not arrested."

WHERE FOUND.

It now seems to be prevalent through nearly all of N. Y., most of the Eastern States and in some parts of Ohio and Pa. In the great California bee-fever a great many diseased stocks were bought up for shipment, by speculators, who were perfectly ignorant of its nature, which resulted in spreading it through that State. Loud complaints came back against unprincipled bee-keepers, attributing blame where it did not always belong. Whether it has increased or diminished within the last year or two, I have not been informed.

WHEN FIRST DISCOVERED.

My first experience will probably go back to a date beyond that of many others. It is now thirty years ago since I noticed the first case. I had kept bees but four or five years when I discovered it in one of my best stocks. It cast no swarm through the summer, and in September, instead of being crowded with bees, contained very few, so few that I dared not attempt to winter it. What was the matter? I had then never dreamed of ascertaining the condition of a stock while there were bees in the way, but was like the unskilfull physician, who is obliged to wait for the death of his patient, that he may dissect and discover the cause. I accordingly consigned the few remaining bees to the "brimstone pit."

DESCRIPTION.

A post-mortem examination revealed the following state of things. Nine-tenths of the breeding-cells contained young bees in the larva state, stretched out at full length, sealed over, dead, black, putrid, and emitting a disagreeable smell. Here was one link in the chain of cause and effect. I learned why there was a scarcity of bees in the hive. What should have constituted their increase, had died in the cells; none were removed, consequently but few cells, where any bees could be matured, were left. But when I attempted to discover the next link in the chain, viz:—What caused the death of this brood, just at this stage of development—I was obliged to stop. Not the least satisfaction could be obtained. All inquiries among the bee-keepers of my acquaintance were met with profound ignorance. They had "never heard of it!" No work on bees that I consulted ever mentioned it.

Subsequently I found more stocks in the same condition. I learned whenever the disease existed to any extent, that the few bees matured were insufficient to replace those that were lost; that the colony rapidly declined, and *never afterwards cast a swarm.*

REMEDIES ATTEMPTED.

I tried pruning out all the combs containing brood, leaving only such as contained honey, and let the bees construct new for breeding. It was of no use; these new combs were invariably filled with diseased brood. The only effectual remedy was to drive out the bees into an empty hive. In this way, when done in season, I generally succeeded in raising a healthy stock. But here was a loss of all surplus honey, and a swarm or two that might have been obtained from a healthy one.

SUPPOSED CAUSE.

I had so many cases of the kind, that I became alarmed, and made inquiry through the agricultural papers for a cause and remedy, offering a reward "for one that would not fail when thoroughly tested." Mr. Weeks, in answer, said that "cold weather in spring, chilling the brood was the cause." (This was several years prior to his article spoken of.) Another gentleman said, that "the accumulation of dead bees and filth during winter, when suffered to remain during spring, was the cause." A few years after another correspondent appeared in one of the papers, giving particulars of his experience, proving very conclusively to himself, and many others, that it was to be attributed to *cold*. Having mislaid the paper containing his article, I will endeavor to quote correctly from memory. He had "three swarms issue in one day; the weather during the day changed from very hot to the other extreme, producing frost in many places the next morning. These swarms had left but few bees in the old stocks, and the cold forced them up among the combs for mutual warmth; the brood near the bottom, thus left without bees to protect it by their animal heat, became chilled, and the consequence was diseased larvæ." He then reasoned thus, "if the eggs of a fowl, at any time near the end of incubation, become chilled from any cause, it stops all further developments. Bees are developed by continued heat, on the same principle, and a chill produces the same effect. Afterwards, other swarms issued under precisely similar circumstances; but these old stocks were covered with a blanket through the night, which enabled the bees to keep at the bottom of the hive. In a few days, enough were hatched to render this trouble unnecessary. These last remained healthy."

He further says, that "last spring was the first time I ever knew them to become diseased before swarming had thinned the population. The weather was remarkably pleasant through April. The bees obtained great quantities of pollen and honey, and by this means, extended their brood further than usual at this season. Subsequent chilly weather in May, caused the bees to desert a portion of brood, which was destroyed by the chill."

This is reasoning from cause to effect very consistently. Had I no experience further than this, I should perhaps rest satisfied as to the cause, and should endeavor to apply the remedy. Several other articles have appeared in different papers, on this subject, and nearly all who assign a cause have given this as the most probable. One says that wintering in the house, and then suddenly transferring to the open air, chills the brood. Now I have known the bee in a chrysalis state, in a few stocks, to be chilled and destroyed by a sudden turn of cold weather, yet these were removed by the bees soon after, and the stocks remained healthy. To me, the cause assigned seems inadequate to produce all the observed results. After close, patient observation for thirty years, I have never yet been wholly satisfied that any one instance of diseased brood among my bees, was thus produced.

We are all familiar to some extent with the contagious diseases of the human family, such as small-pox, measles, etc., and their rapid spread from a given point. We must admit that some cause or causes adequate to the effect, must have produced the first case. To contagion, then, I would attribute the spread of this disease of our bees, in nineteen cases in twenty. I will admit that one stock in twenty or fifty may be affected by a chill to some extent. It is only a portion of brood that is in danger. Only such as have been sealed over, and have not progressed to the chrysalis state are attacked. How many then can there be in a hive, at any one time in just the right stage of development to receive the fatal chill? Of course there will be some, but they will be confined to the cells near the bottom, where the bees have left them exposed. This small number would never seriously damage the stock, if the disease did not spread. Why does it, then, when thoroughly started, spread so rapidly through all the combs in the hive? Will it be said that the chill is repeated every few days through the summer? Or must it not be admitted that something else may continue it? I think that in most cases, there must be other causes, besides the chill, to even originate it.

As our practice will be in accordance with the view we take in this matter, and the result will be somewhat important, I will give some of the reasons that have led to this conclusion.

Once in the month of March, all the bees of a good swarm left the hive and united with another good stock, making double the number of bees at this season; enough to keep the brood sufficiently warm at any time, if other stocks with half or quarter of the number could do so. By the middle of June, the bees were much reduced, and had not cast a swarm. The hive was examined, and the brood found badly diseased.

My best and most populous stocks in spring, are just as liable to be found in this condition, and I might add more so, than smaller or weaker families. I have united two large swarms, and found them diseased the next autumn. (It is probable that they obtained diseased honey.) These cases prove strongly, if not exclusively, that animal heat is not the only requisite. The facts, that when I had pruned out all affected comb from a diseased stock, and left honey in the top and outside pieces, and the bees constructed new for breeding, and the brood in such were invariably affected, slightly at first, and increasing as the combs were extended, led me to suppose that it was a contagious disease, and the poison was contained in the honey. Some of it being left in the hives, the bees had probably fed it to the brood. To test this theory still further, I drove all the bees from such diseased stocks, strained the honey, and fed it to several young healthy swarms soon after being hived. When examined a few weeks after, *every one, without an exception, had caught the contagion.* Here then is a clue to the cause of the spread of the disease, whether we have its origin or not. We will now see if there is any consistency in the theory that it can be transferred from one stock to another.

MANNER OF SPREADING.

Suppose one stock has caught the infection, and but a small portion of the brood is dead. In the heat of the hive, it soon becomes putrid; adjoining cells containing larvæ of the right age, are soon in the same condition. All the breeding combs in the hive become one putrid mass, with an exception, perhaps of one cell in ten, twenty, or a hundred that may perfect a bee. Thus the increase of bees is not enough to replace the old ones that are continually dying off. It is plain, therefore, that this stock must soon dwindle to a very small family. Let a scarcity of honey now occur in the fields, this poor stock cannot be properly guarded, and is easily plundered of all its contents. Honey is taken that is in close proximity to dead bodies, corrupting by thousands, creating a pestilential vapor, of which it has probably absorbed a portion. The seeds of destruction are by this means carried into healthy stocks. In a short time, these in turn fall victims to the scourge, and soon dwindle away, when some other strong stock is able to carry off *their* stores, and this destruction will only cease, perhaps, with the last colony of the apiary. The moth is ever ready with her burden of eggs, which she now deposits without hindrance, directly on the combs. In a short time the worms finish the business, and are pronounced guilty of all the charges, merely because they are found carrying out effects that speedily follow such causes.

In the summer of 1856 there was an extraordinary yield of honey. The old stocks were examined at the usual time after the issue of the first swarms, and the ordinary amount of the disease was found. On the next examination, in September, more than one half the old stocks had become affected, while all the new swarms in the same yard were entirely free. Had these caught the contagion by robbing, the swarms would have participated, and would have been equally affected. A new cause was evidently to be sought for.

MR. WAGNER'S VIEW.

Mr. Wagner, whom I have before mentioned, offered a solution, in substance as follows. After the old queen had left the stock, and all brood had been fed and scaled up, no more food was required for the young bees, till the young queen had laid eggs which should hatch and need it, a period of some two or three weeks. During this interval, the workers continued to collect pollen, the principal food of the larvæ, and store it in the hive. During the time this pollen remained stored in the hive, and before needed for brood, it became soured, decomposed, or by some other process, rendered unhealthy, when fed to the larvæ, and some of them sickened and died. The reason that the swarms were not affected, was, that they had an old queen steadily producing brood, which required food every day, and consumed this pollen in its fresh state, consequently remained healthy. Now here was a theory explaining the origin of the phenomena better than most others, though not perfectly satisfactory. A year or two later, we had a very poor season for swarming, and but few issued. This furnished a sort of test of this theory. If on examination of a given number in the fall, there should be more diseased ones among those that had swarmed, than among those that had not, it would be strong presumptive evidence in its favor. The result was so nearly equal, that it proved nothing. Adding still another hypothesis, and supposing that this poisonous material is not obtained every year, would help it some, yet it would be necessary then to admit that these bad qualities would affect the hive for several years afterward. It would also suggest that the material was not pollen, but some other substance. For instance, the secretions of the Aphis, or Plant Louse. We all know that they are more numerous some seasons than others. Mr. Wagner says that in Germany the Aphides appeared one season, in **countless** myriads, secreting the saccharine fluid in abundance. The bees appropriated large quantities, and as a result, nine-tenths became badly diseased. He suggested that our evergreens produced something of this kind. This is possible, but not probable. Unless our evergreens are different from others, or produce a different race of insects from those produced where the disease never is found, we shall need to look further for a solution.

There is no point connected with bee-keeping, on which I have bestowed so much anxious thought, with such unsatisfactory results. It is very difficult to detect the first hundred or two larvæ that die in a stock. But when nine-tenths of the breeding-cells hold putrid larvæ, there is but little difficulty in making out a diagnosis. The bees are few and inactive. When passing the hive, our olfactories are saluted with nauseous effluvia, arising from this corrupting mass. Now, if we wish, or expect to escape the most severe penalty, our neglect must never allow this stage of progress before such a stock is removed. Therefore, we must watch symptoms, and ascertain the presence of the disease at the *earliest possible moment*.

CAUTION.

As no part of the breeding season is exempt, the stocks should be carefully observed during spring, and early part of summer, with reference to increase of bees. When any are much behind others in this respect, make an examination immediately. The movable comb hive is readily examined by lifting out the combs, but the box-hive must be inverted, and the bees smoked out of the way.

EXAMINATION.

Attention must be directed to the breeding-cells; with a sharp pointed knife, proceed to cut off the ends of some that appear to be the oldest, bearing in mind that young bees are always white, until some time after they assume the chrysalis form. Therefore if a larvæ is found of a dark color, it is dead. Should a dozen or two such be found, the stock should be condemned at once, and all the bees driven into an *empty* hive. On no consideration put them into empty combs, as they would be likely to keep some of the honey for their brood. If it is desirable to put them in a hive containing comb, they may be transferred to it after they have been in an empty one long enough to consume all the honey they have carried with them. (Directions giving for driving in CHAP. XIII.) If honey is scarce at the time, they should be fed. But if it is discovered too late for honey to be collected, it will hardly pay to feed them.

The honey from the old hive may be used, if the poison is first destroyed. This may be done by scalding. Add a quart of water to about ten pounds of honey, stir it well, heat it to the boiling point, and carefully remove all the scum.

Stocks, in which the disease has not progressed too far, will generally swarm. Three weeks after the first swarm, is the proper time to examine them. I make it a rule to inspect all my stocks at this period. It is easily done, as about all the healthy brood, except drones, should be matured in that time. By perseverance in these rules, I allow no stocks to dwindle away until they are plundered by others.

If all bee-keepers were equally careful this disease would only occasionally be found. This is like a careless farmer, allowing a noxious weed to mature seeds, to be wafted by winds to the lands of a careful neighbor, who must fortify himself to continual vigilance, or endure a foul pest. So with a successful apiarian, in sections where it has not appeared, he must be continually on the watch. Vigilance is the price of success.

Again, after the breeding season is over, in the fall,

every stock should be thoroughly inspected, and all diseased ones condemned for stock hives. Even if it should take the last one, it would pay to procure healthy ones instead. Persons wishing to eat the honey from such hives, will experience no bad effects from it, if they are careful to remove the brood combs, as they take it out of the hive.

Careless bee-keepers, when their hives are robbed, feel regret, or are more often vexed with some one, at the result of their own carelessness. The real cause of complaint more often belongs to the owners of the robbing bees, as the honey obtained in this way, probably carries with it more mischief than can be eradicated in a twelve month.

ASSUMED KNOWLEDGE.

It is interesting to read the descriptions of this disease by the would-be bee-doctors, who have never had a case during their "long" experience. They have heard of it, somewhere, and forthwith they know all about it, prescribe remedies, and recommend antidotes. An article appeared in an agricultural paper not long since, with alarming features. After describing the disease, he gave as the only safe remedy, burning the hive, killing the bees, and burying the remainder of the contents; proving that he *knew* nothing of the subject, and had copied from some unreliable source. A person, who will advise such waste, should not be accepted as a teacher of the people. To say he advised it ignorantly, without due consideration, does not help the matter. Why did he assume to teach what he knew nothing about? What is the use of killing a colony of bees, when, if attended to in season, they may be converted into a good stock, worth several dollars? Such hives often contain several pounds of beautiful honey,—why bury it? And why waste one or two pounds of good wax which may be readily exchanged for gold?

He likewise cautioned purchasers of Italian queens, who live in districts where the disease has not appeared, to never procure them from a section where it exists; because if the little combs that are sent with the queen should contain any honey from such a hive, "the disease would go with it, as sure as fate." I have never known such a result in a single instance. Neither have I ever found an experienced apiarian, one who *knew* what he was saying, who advanced such an idea. Should a full colony, badly affected, be sent into a district where it never appeared, there is no need of its being extended. If the bees are simply transferred to an empty hive, and the contents secured from pillage, it would go no further. There would be no loss, except in transferring.

CHAPTER XV.

ANGER OF BEES.

CAUSES OF IRRITABILITY.

Keeping bees good natured, offers a pretty fair subject for ridicule, for it seems rather too absurd to talk of teaching a *bee* anything. Nevertheless, it is worth while to think of it a little. Most of us know, that by injudicious training, horses, cattle, dogs, etc., may be rendered extremely vicious. If there is no perceptible analogy between them and bees, experience proves that they too, may be made ten times more irritable than they are naturally.

Nature has provided them with weapons to defend their stores, and combativeness sufficient to use them when necessary. If they were powerless to repel an enemy, there are a thousand lazy depredators, man not excepted, who would prey upon the fruits of their industry, leaving them to starve. Had it been so arranged, this industrious insect would probably have long since become extinct.

In seasons when buckwheat abounds, they seem to manifest more than usual irritability during its bloom. As soon as a stock is pretty well supplied with this world's goods, like some bipeds, they become haughty, aristocratic and insolent. A great many things are construed into insults, that, in their days of adversity, would pass unnoticed; but now it is becoming and due to their honor to show "a just resentment." It behooves us, therefore, to ascertain what are considered as insults. First, all quick motions about them, such as running, striking, etc., are noticed. If our movements among them are slow, cautious, and respectful, we are often let to pass unmolested, having manifested a becoming deportment. Yet the exhalations from some persons appear to be very offensive, as they attack some much sooner than others, though I apprehend there is not so great a difference as many suppose. Whenever an attack is made, and a sting follows, the venom thus imparted to the air is perceived by others at some distance, who will immediately approach the scene, and more stings are likely to follow.

The breathing of a person into the hive, or among them when clustered outside, is considered in the tribunals of their insect wisdom, as the greatest indignity. A sudden jar, sometimes made by carelessly turning up the hive is another. After being once thoroughly irritated in this way, they remember it a long time, and are continually on the alert; the moment the hive is touched they are ready to salute a person's face. When slides of tin or zinc are used to cut off the communication between the hives and boxes, some of the bees are apt to be crushed, or cut in two. This they remember and retaliate as occasion offers; and it may be when quietly walking in the apiary.

HOW THEY MAKE AN ATTACK.

I must disagree with any one who says that we always have warning before being stung. Two-thirds of them sting without giving the least intimation. At other times, when fully determined on vengeance, I have had them strike my hat, and remain a moment endeavoring to effect their object. In this case, I have warning to hold down my face to protect it from a second attempt which is quite sure to follow. As they fly horizontally, the face held in that position is not so liable to be attacked.*

When they are not so thoroughly angry, they often approach in merely a threatening attitude, buzzing around very provokingly for several minutes in close proximity to one's ears and face, apparently to ascertain our intentions. If nothing hostile or displeasing is perceived, they will generally leave; but should a quick motion, or disagreeable breath offend them, the dreaded result is not long delayed. Too many people are apt to construe these threatening manifestations into positive intentions to sting.

NEVER IRRITABLE WHEN AFTER HONEY.

They never make an attack while in quest of honey, or on their return, until they have entered the hive. It is only in the hive and its vicinity that we may expect them to manifest this irascible temperament. It may be subdued in a great measure, if not entirely, by working quietly and using smoke. Any person, having the care of bees, should be armed with this powerful weapon. As bees are not much affected with smoke while flying in the air, but will there have their own way, we must teach them a proper deportment in the hive.

---

* Striking them down renders them ten times more furious. Not in the least daunted they return to the attack. Not the least show of fear is perceived. Even after losing their sting, they obstinately refuse to desist. The best way is to walk as quietly as possible to the shelter of some bush or to the house. They will seldom go inside of the door.

Those who are accustomed to smoking will find a pipe or cigar very convenient here. But I would not advise any one to make this an excuse for forming a bad habit.*

SMOKER DESCRIBED.

Get a tin tube five-eighths of an inch diameter, five or six inches in length; make stoppers of wood to fit each end, two and a half or three inches long, tapered at the ends. With a nail-gimlet make a hole through them lengthwise; when put together, it should be about ten inches in length. On one end make a notch, that it may be held with the teeth, which is the most convenient way, as you will often want to use both hands. When ready to operate, fill the tube with tobacco, ignite it, and put in the stoppers; by blowing through it, you keep the tobacco burning, while the smoke issues at the other end. This requires blowing almost constantly to keep it burning. I have another mode of using tobacco which is very convenient. Take a piece of old cotton or linen cloth eighteen inches long by six wide. Spread over it a layer of tobacco one-fourth of an inch thick. Roll up and fasten with a needle and thread. Light one end and it will continue burning as long as required. The smoke of decayed wood, commonly known as spunk or touch-wood, is also useful— but there are cases of extreme irritability, where it does not seem to be as efficient as tobacco. We can now subdue these combative propensities, or render them harmless; turn their anger to submission, and force them to yield their treasures to the hands of the spoiler without an effort of resistance. When once overpowered, they seem to lose all knowledge of their power, and no slave can be more submissive.

---

* I continued the practice for years for the mere convenience of the smoke when operating among my bees. But by using some simple substitutes which I will describe, I have managed bees for seven years without pipe or cigar, and much more to my satisfaction.

ITALIANS LESS DOCILE.

After the effect of the smoke has passed off, their former animosity will return. Should any resentment be shown on raising a hive, blow in the smoke; they will immediately retreat, "begging pardon." The Italians, after apparent submission, will return to the attack several times. It often requires one to smoke them, while another operates.

If you wish to take off a box, raise it just enough to blow the smoke under; you can replace it with another without trouble, and a little smoke will keep the bees out of the way. Those in the box are all submission; the box can be carried away and handled as you please, without their becoming irritated, until they once more get home, and then are much more amiable than if the box had been taken without the smoke. They do not seem to realize anything concerning the transaction.

When bees are to be transferred to a new hive, it is unnecessary to be so very particular about the escape of a single bee; no fears need be felt of such as get out. In driving, the loud humming indicates their fear; the upper hive can then be raised safely. After being thus driven out, they may be pushed about with impunity, and will still remain quiet. In short, the use of smoke on all occasions where they would be likely to be disturbed by our meddling with them, has a tendency to keep dormant their combative propensities. When these have never been aroused, there is much less danger of their attacks, while walking or looking among them.

BEE-CHARMS UNRELIABLE.

As for advertised "Bee-Charms" I would recommend very moderate investments until they have been tested. And you will soon have "enough and to spare," unless you get a different article from any I have ever seen.

STING.

The sting of the bee, as it appears to the naked eye, is a tiny instrument of war, so small, indeed, that its wound would pass unheeded by all the larger animals, were it not for the poison introduced at the same instant.

It has been described as being "composed of three parts, a sheath and two darts. Both the darts are furnished with small points or barbs like a fish-hook," that hold it when thrust into the flesh; the bee being compelled to leave it behind.

DOES ITS LOSS PROVE FATAL?

It is said that "to the bee itself this mutilation proves fatal." This is another assertion so often repeated, that perhaps we might as well admit it; as it would be difficult to disprove it. Think of the impossibility of keeping our eye, for five minutes, on a bee that is flying about, after it has left its sting. Yet there are some persons, so very particular about what they accept as *fact*, that they would require that a bee should be watched till it died, before they could be *positively* sure that the loss of its sting caused its death. They might reason from analogy, and say that other insects possess so little sensation that they have been known to recover, after much more extensive mutilation—that beetles have lived for months under circumstances that would instantly kill some of the higher animals—that spiders often reproduce a leg, and even lobsters sometimes replace a lost claw, etc.

I have endeavored to show that there is no great reason for fear in our operations among bees, yet it is idle to suppose that all will manage successfully without some means of defence, especially when dealing with the Italians. The face and hands being most exposed, need some protection. Thick woolen mittens or rubber gloves are best; the sting is generally left when thrust into a leather glove.

PROTECTION.

To protect the face, procure one and a half yards of thin muslin or calico, sew the ends together, and gather one edge on a rubber cord to fit the crown of a hat; cut out an arm hole on each side, and put a string in the bottom to gather it close to the body, or make it shorter and tie around the neck.

Fig. 26.—BEE HAT.

As I do not expect you to work in the dark, we will have a piece cut out in front, and coarse lace, or fine wire-cloth inserted. That which is just fine enough to prevent a bee from passing, is best, as it gives a better chance to see. To keep the lace from falling against the face, sew a wire around it. To facilitate smoking, I have a tube of some convenient material, several inches in length, passing through the lace or wire-cloth, one end of which can be taken in the mouth, and with which the smoke can be directed wherever desired.

Whenever only a partial protection is necessary, a handkerchief is suitable, it is always at hand, and can be put on in a moment. Throw it over the head, letting it fall around the neck and shoulders, covering all but the face. The hat can be put on over it.

REMEDIES FOR STINGS.

It is difficult to tell which are the best remedies for stings. There is so much difference in the effect upon different individuals, and upon different parts of the body, as well as in the depth a sting reaches, that remedies effectual in one instance, will be virueless in another.

For a number of years, I have used none whatever for myself, and the effect is no worse, nor even as bad as formerly. (It is said that this is because the system is hardened to the effects of the poison.) Among the remedies recommended, are saleratus and water, salt and water, soft-soap and salt, a raw onion cut in two and one-half applied, mud or clay mixed wet and changed often, tobacco wet and thoroughly rubbed to get the strength, and constant applications of cold water. To allay the smarting, the application of tobacco is strongly urged, and cold water is spoken of with equal favor to prevent the swelling.

When stung in the throat, drinking often of salt and water is said to prevent serious consequences.

Whether any of these remedies are applied or not, it is hardly necessary to say that the sting should be pulled out as soon as practicable.

CHAPTER XVI.

ENEMIES OF BEES.

Among the enemies of bees, are included rats, mice, birds, toads and insects. But some of these are probably not guilty of any actual mischief. I strongly suspect that the spirit of destructiveness is altogether too active in many people. There are some farmers, so short sighted and vindictive, that, were it in their power, they would destroy a whole class of birds, because some of them had picked a few cherries, or dug out a few hills of corn, when, at the same time, they are indebted to their activity in devouring worms, insects, etc., that would otherwise have destroyed entire crops. It will be well, therefore, to see if we are to be losers or gainers by an indiscriminate slaughter, before we pass sentence on these reputed pests.

RATS AND MICE.

Rats and mice are never troublesome, except in cold weather. The entrances of all hives standing out, are much too small to admit a rat. No damage need be apprehended from them except when the hives are in the house. They appear to be fond of honey, and when it is accessible, will eat several pounds in a short time.

Mice will often enter the hive when on the stand, and make extensive depredations. Sometimes, after cutting a space in the combs, they will make their nests there. The animal heat created by the bees, will make a snug warm place for winter quarters. The "deer mouse" seems to be particularly fond of the bees, while those belonging to the house, appear to relish the honey. Whether they take live bees, or only such as are already dead, I cannot say. Only a part of the bee is eaten, and judging from the fragments left, they must consume quite a number. Whether they take bees or honey, a little care to prevent their depredations, is well worthy of bestowal. As rats and mice have so long been condemned and sentenced as universal plagues, without any redeeming traits, I will say nothing in their favor, and am perfectly willing that they shall be hanged until dead.

A WORD FOR THE KING-BIRD.

But for some of the birds accused of preying upon bees, I would say a word. The king-bird stands at the head of the list of feathered depredators. With a fair trial he will be found guilty, though not so heinously criminal as many suppose. I think we shall find him guilty of taking only drones. In the afternoon of a fair day, he may be seen perched upon some dry branch of a shrub or tree near the apiary, watching for his victims. I have shot him, and examined his crop, after seeing him devour a goodly number, but in every instance the bees were so crushed that it was impossible to distinguish workers from drones. We are told of great numbers of workers being thus found. It may be so, or it may be thus represented by prejudice. The brutal desire of taking life is so strong with some, that a morbid antipathy is allowed to take the place of justice, and a proper defence is not allowed in cases where the suffering party has not the power to enforce it. If the king-bird devoured workers instead of drones, why does he not visit the apiary long before noon, and fill his crop with them? But instead, he waits until afternoon; if no drones are flying, he watches quietly till one appears, although workers may be out by hundreds. If it is asked how they distinguish them, I would suggest that instinct, which teaches most animals the proper kinds of food, might direct the birds in this case.

CHICKENS WILL EAT DRONES.

I have seen chickens which would stand by the hive and devour every drone, as soon as he touched the board, while workers would pass in scores untouched. Whether this loss of the drones is a disadvantage or otherwise, depends entirely upon circumstances. If there is a scarcity of honey, the fewer drones the better. It is a matter of so little importance to the bees, that it would probably not pay for powder to shoot the depredators.

Martins, and a kind of swallow, are said to be guilty of taking bees on some occasions, but as they pursue them on the wing, the remarks concerning the king-bird are applicable to them.

CAT-BIRD ACQUITTED.

The cat-bird also comes in for a share of censure. It is said "he will get right down by the hive, and pick up bees by the hundred." Yet, in the face of this charge, I am disposed to acquit him. With the closest observation,

I find him picking up only young and immature bees, such as are thrown out from the combs. They may be seen about the apiary, as soon as the first rays of light make objects visible, looking for their morning supply, as well as frequently during the day. Should an unlucky worm be in sight just then, looking up a place to spin a cocoon, or a moth be reposing on some corner of the hive, its fate is at once decided. Before destroying this bird, it would be well to judge from actual observation of the truth of the charges against him, else we may "destroy a friend instead of a foe."

THE TOAD.

A toad is discovered near the hives, and is forthwith executed as a bee-eater. Says one—"he ought to be killed for his looks, if nothing else." He is thus often sacrificed really on account of his appearance, on the nominal pretence that he is a villain. After he is despatched, the complaint is made that the bugs that he might have destroyed, "have eaten up all the little cucumbers and cabbages!" His food is probably small insects. I had strong doubts of his being a bee-eater for a long time after the first editions of this work were published, notwithstanding the positive assurance by some of my bee-keeping friends, that he was guilty. I watched closely for years, without discovering anything to confirm the assertion. At last, one dark, cloudy day, when but few bees were stirring, I found a corpulent fellow perched upon a stand, close to the entrance, seeming very much at home, and gazing at nothing with the most stupid indifference. A bee lit beside him, when after a slight motion of the eye, his mouth opened, and closed like a flash of lightning. The bee was gone!—his long flexible tongue had deposited it beyond the reach of help.

Again he was motionless, and mute as a stone, waiting

for another victim with the most provoking complacency. I found in his stomach more than a dozen which he had already swallowed; they were dead, but not mutilated. Notwithstanding his shabby ingratitude in impudently committing the crime before my very face, after fooling me so long with his innocent looks, and allowing me to plead his case for years, I am going to hand him over to the judge with a strong "recommendation to mercy." I trust he will reform, and not cultivate a taste for beautiful Italians. When he can control his appetite, so as to be content with the delicacies that the garden affords, he may rise in our estimation.

He can be excluded from the apiary, if desired, by a close fence, a foot in height.

BLACK WASP.

But little can be said in favor of the black wasps that visit the hives in the sunny days of spring. They seem to have no other object than to tease and irritate the bees. I never could discover that they entered the hive for purposes of plunder. They have frequent battles with the bees, but I never saw any bees devoured, or carried off, or even killed, although sad havoc is sometimes reported. After the first of June, they are seldom troublesome. The yellow wasps or hornets that are around in autumn, are of but little account; their object is honey which they take when they can, but are not apt to enter the hive among the bees.

ANTS—A WORD IN THEIR FAVOR.

Ants come in for a share of condemnation. These industrious little insects shall have my efforts for a fair hearing. Many bee-keepers are wholly ignorant, most of the time, of the real condition of their stocks. Many causes, independent of ants, induce a reduction of population.

Suppose the bees are so reduced as to leave the combs unprotected, and the ants enter and appropriate some of the honey. The owner comes along just then, and sees them engaged: "Ha, you are the rascals who have destroyed my bees," he exclaims, without a thought of looking for causes beyond present appearances. They are often unjustly accused by the farmer, of injuring his little trees, by causing the tender leaves to curl and wither. Inquiries are often made in agricultural papers for means of destroying them, when the real cause of the mischief is the Aphides, that are upon the leaves and stalks in hundreds, robbing them of their important juices, and secreting a fluid highly prized by the ants.

The habits of the small black ants give rise to suspicion of mischief. They live in communities of thousands, having their nests in old walls, old timber, or in the earth. From these nests a string of ants may be sometimes traced for rods, going after, and returning with food. During wet weather, such as would make the earth, and many other places too damp and cold for a nest, they look out for better quarters. The top or chamber of our bee-hives affords a desirable shelter. The animal heat from the bees renders it perfectly comfortable. How can we blame them for choosing a location so completely supplying all their wants? But the careless observer discovering their train to and fro, from their nest on the hive, exclaims: " Why, I have seen them going in a continual stream to the hive after honey;" when a little scrutiny would show that the nest was on the top of the hive, and they were going elsewhere for food, not one to be seen entering the hive among the bees for honey. When honey is unprotected by bees, and left where they can have access, they will naturally carry off some, but it may be easily secured.

SPIDERS.

Spiders are of course a considerable annoyance to the apiarian, as well as to the bees; not so much on account of the number of bees consumed, as from their habit of spinning a web about the hive, that will occasionally take a moth, but will probably entangle fifty bees the while.* They are probably in fear of the bees, or else, they do not relish the bee as food, as one caught in the morning is frequently untouched during the day. This web is often exactly before the entrance, entangling the bees as they go out and return, irritating and hindering them greatly, though they often escape after repeated struggles. I have removed a web from the same place, every morning for a week, that was renewed at night with astonishing perseverance. The redeeming qualities of the spider are few, and are more than balanced by its evil propensities. Their sagacity will sometimes find a place of concealment, not easily discovered. At the approach of cold weather, the box or chamber of the hive, being a little warmer than other places, will attract a great many there to deposit their eggs. Little piles of webbing may be seen attached to the top of the hive, or sides of boxes. These contain eggs for the next year's brood. This is the time to destroy them, and save trouble for the future.

MOTH.

If we combine into one phalanx all the depredators yet named, and compare their ability for mischief, with that of the wax-moth, we shall find their powers of destruction but feeble in comparison. From the moth herself, we would have nothing to fear were it not for her progeny; a hundred or thousand vile worms, whose food is principally wax or comb.

---

* Not long since, an eminent apiarian recommended the spider as an assistant in destroying the moth.

ENEMIES OF BEES. 235

As the instincts of the flesh-fly direct her to a putrid carcass to deposit her eggs, that her offspring may have their proper food, so the moth seeks the hive containing combs, where the natural food of her progeny is at hand. During the day a rusty brown miller, with wings close to the body, may be often seen lying perfectly motionless, on the corner of a hive, or on the under edge of the top, where it projects over. They are more frequent at the corners than anywhere else, one-third of their length projecting beyond it, appearing much like a sliver on the edge of a board that is somewhat weather-beaten.

Fig. 27.—BEE MOTH—TWO MALES AND ONE FEMALE.

Their color so closely resembles old wood, that I have no doubt their enemies are often deceived, and they thus escape with their lives. As soon as darkness shuts out the view, and there is no danger of their movements being discovered, they throw off their inactivity, and commence searching for a place to deposit their eggs, and woe to the stock that has not bees sufficient to keep them from the comb. Although their larvæ generally has a skin that the bee cannot pierce with its sting, it is not so with the moth, and they seem to be aware of the fact, for whenever a bee approaches, they dart away with a speed ten times greater than that of any bee disposed to follow. They enter the hive, and dodge out in a moment either from fear of the bees, or from having actually encountered them. Now it needs no argument to show, that when our stocks are well protected, there must be a poor chance for depositing eggs upon the combs, which instinct teaches them is the proper place. But they *must* leave them somewhere.

WHERE THEIR EGGS ARE DEPOSITED.

When driven from all the combs within, the next best places are the cracks and flaws about the hive, that are lined with propolis; and the dust and chips that fall on the floor-board of a young swarm not full. This last material is mostly wax, and answers very well instead of comb. The eggs will hatch here, and the worms sometimes ascend to the comb, hence the necessity of keeping the bottom brushed off clean. It will prevent those hatched on the

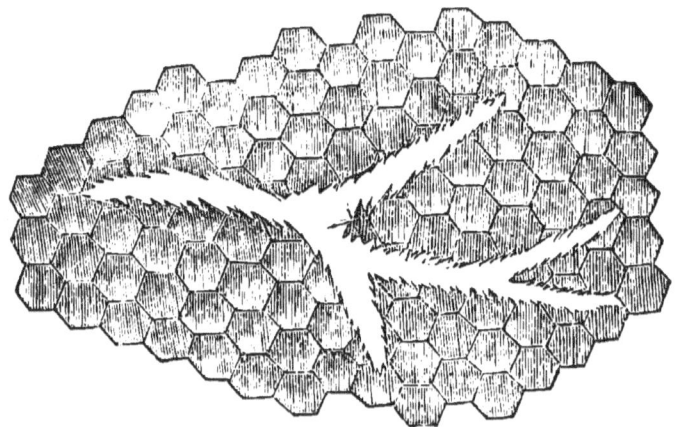

Fig. 19.—WORM GALLERY IN THE COMB.

bottom from going up, also prevent the bees from taking up any eggs on their feet, if this should happen to be the method by which they get among the combs of a populous stock. They are often detected there, and I can conceive of no other means by which they can be deposited. A

ENEMIES OF BEES. 237

worm lodged in the comb makes his way either to the centre, or between the heads of the young bees in the cells and the sealing, and as he proceeds, eats a passage, lining it with a shroud of silk, and gradually enlarging it, as he increases in size. When combs are filled with honey, they work on the surface, eating only the sealing. In very weak families, this silken passage is left untouched, but is removed by all strong colonies. I have found it asserted that "the worms would be all immediately destroyed by the bees were it not for a kind of dread of touching them, until compelled by necessity." As the facts which led to this conclusion are not given, and I can find none confirming it, perhaps I shall be excused for being "of little faith." On the contrary, I find to all appearance, an instinctive antipathy to all intruders, and they immediately remove them when possessing the power.

Fig. 28.—WORM GALLERY REMOVED FROM THE COMB.

WORMS SOMETIMES WORK IN THE CENTRE OF COMBS.

When a worm is in the centre of a comb filled with brood, its passage is not at first discovered. The bees, to get it out, must bite away half the thickness of the comb, removing the brood in one or two rows of cells, sometimes for several inches. This will account for the number of immature bees found on the floor-board at morning, in the spring; as well as in stocks and swarms but partially protected after the swarming season.

BEES MUTILATED BY WEB.

Sometimes a half dozen young bees nearly mature will be removed alive, all webbed together, fastened by legs, wings, etc. All their efforts to break loose prove unavailing. Others may be seen running about with their wings mutilated, part of their legs eaten off, or tied together. These are often the first symptoms of worms at this season. (July and August.) Although unfavorable, it might be worse. It shows that the bees are not yet discouraged, that when they find the worms present, they have still sufficient energy to make an effort to rid themselves of the nuisance. Should the apiarian now give them a little assistance for a few days, they will soon be in a prosperous condition. The hive should be frequently raised, and everything brushed out clean. If a new swarm part full presents these indications, it should be turned over, perhaps, once a week, till the worms are mastered, and the corners inside examined for the cocoons which may be easily detached and destroyed.

In turning over a hive part full, in warm weather, you should first observe the position of the combs, and let the edges rest against the side of the hive, otherwise they may bend, and break loose when the hive is again set up.

When a hive is full of combs, the edges are usually more firmly attached, and it is of less consequence which way it is turned; yet in very warm weather the honey will run out of drone cells, if perpendicular.

BEES FASTENED IN THE CELL.

In *very* small swarms, hundreds of young bees may be frequently seen with their heads out of the cells, endeavoring to escape, but firmly held inside by moth webs. I have known a few instances in such circumstances, where it appeared as if the bees had designedly cut off the whole sheet of comb, and let it drop, thereby ridding themselves of all farther trouble.

DIFFERENT APPEARANCE IN OLD STOCKS.

But when the bees in old stocks make no effort to dislodge the enemy or his works, the case is somewhat desperate. We must look for something entirely different from the foregoing symptoms. But few young bees will be found. In their place we shall find the fœces of the worms dropped on the board. The chips dropped by the bees, in biting off the covering of the cells, to get at the honey, closely resemble them. To detect the difference requires close inspection. The color of the fœces varies with the color of the combs on which the worms feed, from white to brown and black. The size of these grains will vary in proportion to the size of the worm, from a mere speck to nearly as large as a pin head; shape cylindrical, with obtuse ends, length about twice the diameter. By the quantity we can judge of the number of worms. If the hive is full of combs, the lower ends may appear perfect, while the middle or upper part is sometimes a mat of webs.

Whenever our stocks have become reduced from overswarming or other causes, the ravages of worms are to be expected. Here is another important reason for knowing the *actual* condition of our bees at all times; we can detect the operations of the worms very soon after they commence. In some instances we can save the stock by breaking out most of the combs, leaving just enough to be covered by the bees. When success attends this operation, it *must* be performed before the worms have made a permanent lodgment. When the stock is weak, and appearances indicate the presence of many worms, it will be the least trouble in the end, and generally, the safest method, to drive out the bees at once, and secure the honey and wax. The bees may do a little, if put into a new hive, but if they should do nothing, it would be no worse than to leave them in the old hive till the worms had destroyed all, and matured a few thousand moths in addition to those otherwise produced, thereby multiplying the chances of damage to other stocks a thousand fold. It is probably remembered that I said, when bees are removed from a hive in warm weather, that if the hive were not infested with worms at the time, it soon would be, unless smoked with sulphur.

WORMS GROW LARGER WHEN UNDISTURBED.

In a hive thus left without bees, the worms will grow one-half or two two-thirds larger than when their right to the comb is disputed. In one case they often make their growth, and actually wind up in their cocoon, when less than an inch in length; in the other, they will quietly fatten till they are an inch and a half long, and as large as a pipe-stem.

Fig. 1.—MOTH-WORM.

When first hatched from the egg it is difficult to discern them with the naked eye. Their rapidity of growth depends as much, or more, on the temperature in which they are, than upon their good living. A few days of hot weather, may develop the full-grown worm, while it would require weeks and even months in a lower temperature.

The worm, after spinning its cocoon, soon changes into a chrysalis, and remains inactive for several days, when it makes an opening in one end, and crawls out. The time necessary for this transformation is also governed by the temperature, although, I think but few ever pass the winter in this state.

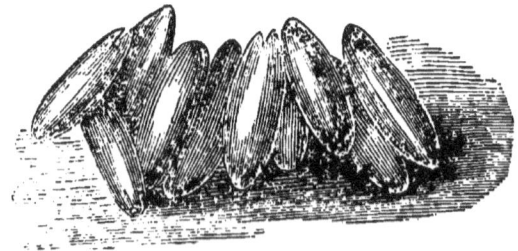

Fig. 29.—COCOONS OF THE MOTH WORM.

A moth will rarely be found before the end of May, and not many are seen till the middle of June; but after this time they are more numerous till the end of the season.

FREEZING DESTROYS THEM.

It is well demonstrated that the moth, its eggs, larvæ, etc., cannot pass the winter without warmth sufficient to prevent freezing. It can be shown thus. Take all the bees out of a hive in the fall, and without disturbing the honey and comb, put it in a cold chamber where it can freeze thoroughly. In the following March, introduce bees, and when not contiguous to a stock containing worms, not a single worm will be produced before the middle of June, or until the eggs of some moth matured in another hive have had time to hatch. Such hives may be kept for swarms, without any appearance of worms.

The discovery that worms and eggs can be frozen to death, has led to a plan by which the whole race of moths can be exterminated from an apiary, and only reappear from those of neighbors. When aided by movable combs, it is entirely practicable. I have tried it to some extent, but since learning that the Italians resisted the moth so much more effectually than the natives, I have not practiced it extensively.

EXTERMINATION OF THE MOTH.

It is simply to expose the combs, free from bees, to a temperature of 18° below freezing, for ten to twenty hours after they are once thoroughly cold. To describe more minutely, I would say that during December or early in January, before the bees have much brood, is the best time.

Take the hive to a dark room, using artificial light to keep the bees from flying. Take out one comb, with honey enough to last for several days, and put it in an empty hive. Set over this, another empty one of the same size,

11

without top or bottom. Take out the next frame, and hold it down in the top hive, and shake or brush the bees into the lower one. As the frames should be put back into the hive in the same relative position, it is well to number them before any are removed.

When the bees have been taken from all the combs in turn, the latter may be put away to freeze. Each comb should be separated from the others, at least several inches, unless they can have plenty of time to freeze.

Combs, close together as they are in the hive, will remain warm a long time. After being sufficiently frozen, they must be warmed for several hours before they are in proper condition to receive the bees. The comb left with the bees must undergo the same process. If preferred, one half the combs may be taken first, and then changed for the other half. In case you have more good colonies than you care to keep, you may kill the bees, freeze the combs, and transfer the colonies into them to remain; it will save transferring once.

The bees, not comprehending what all the shaking is about, will become very indignant at the unnecessary abuse, especially when it is repeated at the second transfer.

Could a cheap freezing mixture of proper intensity be applied for a length of time sufficient to freeze to the centre, the operation might be performed in November, or as soon as all the brood is hatched.

If all the bee-keepers in a neighborhood, town or county, could be induced to do this perseveringly for a year or two, it is plain that the extermination would be so nearly complete, that it would take a long time for the moths to regain their former position.

Any one situated a goodly distance from neighboring bees, would find the summer pretty well advanced before there was even any appearance of moths. What the difference would be, on the average, in the prosperity of our bees, betwen those free from moths, and those, as ordinarily troubled with them, must be guess-work. It is evident that it is sometimes not less than one-fourth or one-third. These remarks are with reference to the black bees. The Italians take care of themselves so well that much trouble in this respect is unnecessary.

SELDOM ENTIRELY EXEMPT IN ORDINARY MANAGEMENT.

But hives in which bees are wintered in the usual way, are seldom or never exempt. It is probably impossible to winter bees without preserving some eggs, or a few worms, at the same time. The perfect moth probably never survives the winter; the only place in which the chrysalis would be safe, I think must be in the vicinity of the bees, and a good stock will never allow it there,—but eggs it would appear are suffered to remain. In the fall, at the approach of cold weather, the bees are apt to leave the ends of the combs exposed. The moth can enter, and deposit her eggs directly upon them; these together with what are carried in by means before suggested, will insure a good supply for the coming season.

The warmth generated by the bees will keep these eggs from freezing. When warm weather approaches in the spring, those nearest the bees probably hatch first, commence their depredations, and are removed by the bees. As the bees increase and occupy more comb, more worms are hatched. In this way, even a small family will hatch and get rid of all the eggs that happen to be in their combs, and not be destroyed. This is the time that the apiarian may be of service in destroying the worms, as they are thrown on the floor by the bees.

In July or August a single moth may enter an exposed hive, and deposit her burden of several hundred eggs as in the other case, but the heat from the bees is now unnecessary to hatch them. The weather at this season will make any part of the hive warm enough to set her whole brood at work at once, and in three weeks all may be destroyed.* This, and the fact that more moths exist now than before, may account for the greater number of stocks destroyed at this season. Yet, it is considered extremely bad management to allow honey or combs to be devoured by this disgusting creature. It is *necessary* to know the condition of the stocks to prevent their getting the start. These duties should be fully considered before we take the responsibility of the care of bees.

The only time when we can rest and feel safe is when we *know that all our stocks are full of bees.* Even the moth-proof hive containing combs will be scented out by the moth when there are no bees to guard it. An argument to show that a moth can go where a bee can, is unnecessary, and a little observation will prove that her eggs sometimes go where *she* is not allowed.

REMEDIES.

But as we cannot always have our bees in proper condition, it is well to adopt some of the means recommended to diminish the number of moths. In July and August it is a good plan to put a few pieces of old dry combs near the hives, in a box or other place, as a decoy, where the moth may have access. She will deposit a great many of her eggs here, instead of in the hive, and they can be easily destroyed. Make it a rule to destroy all the worms that can be found at any time, particularly in spring; likewise, all cocoons. A great many worms can be enticed to web up under a trap of elder, when it is an easy matter to dispatch them. Destroy all the moths that are seen about the hive. They are very much like the flea, "when you put your finger on him, he is not there;" a careful move must be made, else she darts away. Probably the most

---

* Worms create much warmth of themselves.

expeditious mode is to make them drunk. Mix with water just enough molasses and vinegar to make it palatable; put it in saucers or other dishes, and set among the hives at night. Like nobler, if not wiser beings, when once they have tasted the fatal beverage, they seem to lose all power to leave the fascinating cup; and give way to appetite and excitement till a fatal step plunges them into destruction. The next morning finds them yet wallowing in filth, weak and feeble. Whether they would recover from the effects of their carousal, if lifted out of the mire, and carefully nursed like other specimens of creation, I never ascertained. With but little trouble, a chicken or two can be taught to be on hand, and will greedily devour every one. Hundreds may be caught in this way, mixed with many other kinds. I have thought that this liquid answered a better purpose after it had fermented.

CHAPTER XVII.

WAX.

The unreflecting observer, seeing the bees enter the hive with a pellet of pollen on each posterior leg, is very apt to conclude that it must be material for comb, as it does not resemble honey. There is so little thought on the subject that they do not imagine any other use for it. Others suppose that it will change to honey after being stored in the hive a while, and wonder at the curious phenomenon, but when asked how long a time must elapse before it takes place, they cannot tell exactly, but they " have found cells where it began to change, as a portion near the outer end of the cell had become honey, and, no doubt, the remainder would, in time." This conclusion has doubtless arisen from the fact, that cells only about two-thirds full of pollen, are often finished with honey.

WHAT IS IT?

Those who contend that combs are made of pollen, would probably abandon the idea, after seeing the bees belonging to a hive filled to the last inch with comb, collecting and bringing home just as much pollen as those belonging to a hive half full. The question as to where the bee gets wax to construct its combs, is very much like asking where the cow gets her milk, or the ox his tallow. I believe all close observers agree that wax is a secretion natural only to the bee. Honey, and syrup made of sugar are probably the only substances from which they secrete it. From experiments with them, Huber has decided that either of these substances, mixed with a little water, is all sufficient for its production. From experiments of my own, I am satisfied that he is correct, and that pollen is unnecessary. The experiment may be tried by shutting up a swarm when first hived, and feeding them with honey only. A few of the bees will probably have some pollen, though not enough to make a comb three inches square, and to be certain, time must be given them to exhaust it. In three or four days take out the bees, and remove the combs; enclose them again, and feed as before.

Repeat the process, until satisfied that no pollen is needed in the composition of wax. Huber removed the combs five times with the same result at every trial. Whenever bees are confined in hot weather, *air and water are absolutely necessary.*

HOW IT IS OBTAINED.

We will now describe the first appearance of wax and how it is produced. When a swarm of bees is about leaving the parent-stock, three-fourths or more of them will fill their sacs with honey. When located in their new home, of course, no cells exist to hold it, and it must remain in the sac or stomach for several hours. The consequence is, that thin white scales of wax, one sixteenth of an inch in diameter, somewhat circular, are formed between the rings of the abdomen, on the under side. Fig. 30 shows the abdomen of the bee, enlarged, with the scales of wax between the rings. With the claws of one of their hind-legs, one of these is detached and conveyed to the mouth, and then pinched with their forceps or teeth until one edge becomes somewhat rough; it is then applied to the comb being constructed, or to the roof of the hive. The first rudiments of comb, are often to be seen within the first half hour after the swarm is hived. Transferring the swarms to other hives from one to forty-eight hours after being hived, will show their progress.

Fig. 30.

COMMENCEMENT OF A COMB.

I have found that wax is attached to the top of the hive, at first, without the least order, until some of the blocks or lumps are sufficiently advanced for them to begin cells. The scales of wax are welded together, without regard to the shape of the cell, then an excavation is made on one side for the bottom of a cell, and two others on the opposite side, the division between them being opposite the centre of the first. When this piece of comb is an inch or two in length, two other pieces, at nearly equal distances on each side, are commenced. If the swarm is large and honey abundant, it is common for two pieces of comb to be started at one time, on different parts of the top; the sheets in the two places are as often at right angles, as parallel, or any other way just as chance directs them. The little lumps that are deposited at random, at first, are removed as they proceed.

While the combs are in progress, the bases of the cells near the edge are always kept much the thickest, and are worked down to the proper thickness with their *teeth*, and polished smooth as glass. The ends of the cells also, as they lengthen them, will always be found much thicker when finished, than any other part of them.

In the History of Insects, published by Harper, is a minute account of the first foundation of combs, somewhat amusing, if not instructive.

Huber, it is said, "having provided a hive with honey and water, it was resorted to, in crowds, by bees, who having satisfied their appetite, returned to the hive. They formed festoons, remained motionless for twenty-four hours, and after a time scales of wax appeared. An adequate supply of wax for the construction of a comb, having been elaborated, one of them disengaged itself from the centre of the group, and clearing a space about an inch in diameter, at the top of the hive, applied the pincers of one of its legs to its side, detached a scale of wax, and immediately began to mince it with the tongue. During the operation, this organ was made to assume every variety of shape; sometimes it appeared like a trowel, then flattened like a spatula, and at other times like a pencil, ending in a point. The scale, moistered with a frothy liquid, became glutinous, and was drawn out like a ribbon. This bee then attached all the wax it could concoct to the vault of the hive, and went its way. A second now succeeded, and did the like; a third followed, but owing to some blunder, did not put the wax in the same line with its predecessor, upon which, another bee apparently sensible of the defect, removed the displaced wax, and carrying it to the former heap, deposited it there, exactly in the order and direction pointed out."—

Now, I have some criticisms to make on this account. First, in the usual course of swarming, it is unnecessary to provide the honey and water, as they come laden with honey from the parent-hive. Next, to form festoons, and remain motionless twenty-four hours to concoct the wax, is not their custom. They either swallow the honey long enough before leaving home, to have the wax ready, or less time than twenty-four hours is necessary to produce it. I have frequently found lumps, about the size of a pin-head, attached to a branch of a tree where they had clustered, when they had not been there over twenty-five minutes. I have had occasion many times to change the swarm to another tenement, an hour or two after they were hived, and have found places on the top nearly covered with wax. How he managed to see a bee " quit the group," or to ascertain that the tongue was the only instrument used in moulding the scale of wax, is more than I can comprehend. To witness the whole process in all its minutiæ, in this stage of comb making has never been my good fortune, and I am sometimes inclined to doubt the success of others. I have had glass hives and put swarms in them, and always found the first rudiments of comb so entirely covered with bees, as to be unable to see anything of the operation. The only time when I have been able to witness the process, with any degree of satisfaction, has been when the combs approached the glass and there were but few bees in the way, then, with a little patience, some part of the process may be seen.

When two combs approach each other in the middle of the hive at right angles or nearly so, they are not joined; but when at an obtuse angle, the edges are generally united, making a crooked sheet of comb. It is evident, that where the two combs join, there must be sóme irregular cells, unfit for rearing brood.

CROOKED COMBS.

Crooked combs do not seem to affect the prosperity of the hive. Combs built in the Cross Bar or Movable Comb Hive, are usually straight when the under side of each bar is brought to an edge like a knife. But there are exceptions enough to almost annul the rule, in ordinary management. It is found, however, that a smooth sharp edge is followed much better than a rough one. Sometimes, after combs are started straight, the bees will take some other direction, and by the time the combs reach the bottom of the hive, they may be at right angles with their course at the top.

STRAIGHT COMBS.

I recently made the discovery, that if one end of the hive was elevated 30°, straight combs would be the result throughout, especially if the hive were perfectly level the other way.

Sometimes there will be corners and spaces not wide enough for two combs, and too wide for one of the proper thickness for breeding. As bees generally use all their room to the best advantage, a thick comb will be the result, and when used for breeding, the cells are cut down to the proper length.

QUANTITY OF HONEY TAKEN BY A SWARM.

A large swarm will probably carry some five or six pounds of honey from the mother colony. It is impossible to determine the exact amount as the weight of the bees is very uncertain.

"I can tell you," some one exclaims, "I saw some weighed—so many weigh just eight ounces. Are you sure that nothing else was weighed—no honey, bee-bread, fœces, or other substances? "Can't say,—never thought of that." It is important, if we wish to know the weight of bees alone, that we weigh nothing else. It is evident, if a few thousand weigh three pounds, when nothing is in their sacs, that they would weigh several pounds more, when filled with honey. Hence, the fallacy of judging of the size of a swarm by weight, as one swarm might issue with half as much honey as another. Perhaps eight pounds would be a correct average for the weight of bees and honey, in large swarms. This honey whatever it amounts to, cannot be stored, till combs are constructed to hold it. This principle holds good till the hive is full. That is, whenever they have more honey than the combs will hold, and there is room, they will construct more comb. But they seem to go no farther than this in comb-making. However large the swarm may be, this compulsion appears necessary to fill the hive.

MAKING DRONE CELLS.

Drone-cells are seldom made in the top of the hive, but some are generally joined on the worker cells, a little distance from the top; others near the bottom. There seems to be no rule about the number of such cells. Some hives will contain twice as many as others. It may depend on the yield of honey at the time; if plenty, more drone-cells, and *vice versa*. It has been suggested that more drone-cells are built while filling the hive, when the swarm has an old queen. If the hive be very large, no doubt an unprofitable number will be constructed. Where the large and small cells join, there will be some of irregular shape; some with four or five angles. Even where two combs of cells, the same size, join, making a straight comb, they are not always perfect.

SOME WAX WASTED.

When constructing comb, they are constantly wasting wax, either accidentally or voluntarily. The next morning after a swarm is located, the scales may be found, and will continue to increase as long as they are working it; the quantity often amounts to a handful or more. It is the best test of comb-making that I can give. Clean off the board, and look the next morning, you will find the scales in proportion to their progress. Some will be nearly round as at first; others more or less worked up, and a part like fine saw-dust.

WATER NECESSARY.

Whenever bees are engaged in making comb, a supply of water is absolutely necessary. When no pond or stream is within convenient distance, the apiarian will find it economical to place water within their reach. As the necessity for it always occurs in a busy season, it will save much valuable time. It should be so situated, that the bees may obtain it without danger; a barrel or pail has sides so steep, that a great many will slip in and drown. A trough made very shallow, with a broad strip around the edge to afford an alighting place, should be provided. It should contain a float, or a few shavings scattered in the water, with a few small stones laid on them to keep them from blowing away. A tin dish, an inch or two in depth, will answer very well. The quantity needed, may be easily ascertained;—give them just enough, and change it daily. I have no trouble of this kind, as there is a stream of water within a few rods of my hives; but I have an opportunity to observe the number engaged in carrying it. Thousands may be seen in June and August filling their sacs, while a continuous stream of bees is on the wing, going and returning.

CELLS UNIFORM IN SIZE.

The exact and uniform size of their cells is perhaps as great a mystery as anything pertaining to bees; yet we meet the second wonder before we are done with the first.

In building comb, they have no square and compass as a guide; no master mechanic takes the lead, measuring and marking for the workmen; each individual bee is a finished mechanic! No time is lost in apprenticeship, no service given in return for instruction. Each is accomplished from birth! What one begins, a dozen may unite to finish! Each specimen of their work may be taken as a model! He, who arranged the Universe, was their instructor. Yes, a profound geometrician planned the first cell, and knowing what would be their wants, implanted in the sensorium of the first bee, an instinctive knowledge of all things necessary to its welfare, which remains unimpaired in its latest descendant.

They need no lectures on domestic economy to tell them that the use of the base of one set of cells, on one side of the comb, for the base of those on the opposite side, will save both labor and wax; no mathematician, that a pyramidal base, with just three angles, and just such an inclination, is the exact shape needed, and will take much less wax than if round or square—that the three-angled base of one cell, forms a part of the base of three other cells on the opposite side of the comb—that each of the six sides of one cell, forms one side of six others—that these angles and these only would answer the ends required.

"The bees appear," says Reaumur, "to have a problem to solve, which would puzzle many a mathematician. A quantity of matter, being given, it is required to form out of it cells, which shall be equal and similar, and of a determinate size, but the largest possible with relation to the matter employed, while they shall occupy the least possible space."

How little does the epicure heed when feasting on the fruits of their industry, that each morsel tasted, must destroy the most perfect specimens of workmanship! That in a moment he demolishes what it has taken hours, yes, days and weeks of assidious toil and labor to accomplish.

MELTING OF COMBS.

When extreme hot weather occurs immediately after the bees have been gathering from a plentiful harvest for two or three weeks, or even during the yield, the wax composing new combs is very liable to be softened, till they break loose from their fastenings and settle to the bottom. The first indications of such an accident, when the hive is half or two-thirds full, are: clusters of bees on the outside, and honey running out at the bottom.

Sometimes the injury is trifling, only a piece or two slipping down; at other times the whole contents fall in a confused and broken mass, the weight pressing out the honey, and besmearing the bees, which, being thus soiled, creep out and away from the hive in every direction.

I once had some stocks ruined, and others injured in this way, by hot weather, about the first of September, immediately after the buckwheat season. The bees, or most of them, being covered with honey, together with what ran out of the hive, at once attracted others to the spot, who carried off the entire contents in a few hours. This was an uncommon occurrence; I have known but one season in twenty-five years, when it occurred after the failure of honey from the flowers. It usually happens during a copious yield, and then other bees are not apt to be troublesome by robbing.

To prevent such mishaps, ventilate by raising the hives on little blocks at the corners, and *effectually protect them from the sun;* if necessary, wet the outside of the hive with cold water. After the loss of those before mentioned, I kept the rest of the new hives wet, through the middle of the day, and I have no doubt but that I saved several by the means. I had some trouble with those in which only a piece or two had fallen, and started just honey enough to attract robbers. It was not safe to close the hive to exclude them, as this would have increased the heat, and proved certain destruction.

The best protection I found was a few stems of asparagus around the bottom of the hive; this permitted a circulation of air, and at the same time made it very difficult for the robbers to approach the entrance, without creeping through this hedge, and with this assistance the bees of the hive defended themselves, till all wasting honey was taken up.

When the hive is nearly full, and but one or two sheets come down, their lower edges will rest on the floor, and the other combs will keep them in an upright position, until fastened by the bees. It is generally best to leave such pieces as they are. If the hive is but half full or little more, and such pieces are not kept perpendicular by the remaining combs, they are apt to be broken and crushed badly, by falling so far, and most of the honey will be wasted. To save this, it must be removed, unless it can be caught in a dish. Be careful not to turn the hive on its side, and thus break the remaining combs. Such combs as contain brood and little honey, may be left for the brood to mature. Should the bees be able to take up the honey without much waste occurring, it would be advisable to leave it; it would assist greatly in filling up. But these broken pieces should be removed before they interfere with the extension of the other combs. A part of the bees is generally destroyed, but the majority will escape; even such as are covered with honey, if they are not crushed, will clean it off, and soon be in working order, when others do not too officiously assist in removing it. An ample yield of honey is the best protection against this disposition to pillage. After the first year, combs become thicker, and are not so liable to give way.

CHAPTER XVIII.

PROPOLIS.

The origin of propolis is a subject upon which apiarians fail to agree. It is asserted by some that, when the bees need it, they always have it, and therefore they contend that it is elaborated like wax; while others believe it to be a resinous gum, exuding from certain trees, and collected by the bees, in the same manner as pollen.

HOW OBTAINED.

Huber tells us that, "near the outlet of one of his hives he placed some of the branches of the poplar, which exudes a transparent garnet-colored juice. Several workers were soon seen perched upon these branches; having detached some of this resinous gum, they formed it into pellets, and deposited them in the baskets of their thighs; thus loaded, they flew to the hive, where some of their fellow-laborers instantly came to assist them in detaching this viscid substance from their baskets."

I am convinced that it is a natural secretion of some kinds of trees, as I have seen the bees collecting it, and have frequently seen them enter the hive with what appeared to be the pure article on their legs, resembling pollen, except that the surface is smooth and glossy. It is of much lighter color when new, than it is after it gets a little age.

HOW DISCHARGED.

I have also seen bees through the glass, when they seemed unable to dislodge it themselves, and were continually running around among those engaged in soldering and plastering. When one required a little, it seized the pellet with its forceps, and detached a portion. The whole lump will not cleave off at once, but firmly adhere to the leg, and from its tenacity, a string an inch long, will sometimes be formed in separating. The piece obtained is immediately applied to their work, and the bee is ready to supply another with a portion. It doubtless gets rid of its whole load in this way, but it is difficult to watch the whole process, as the bee is soon lost among its fellows.

The buds of many trees are protected from the elements, by a kind of gum or resinous coating. It may be found in many species of *Populus*, particularly the Balsam Poplar (*Populus balsamifera*) and the variety (*candicans*) known as the Balm of Gilead. By boiling the buds of these trees an aromatic gum or resin may be obtained, the odor of which is very similar to that emitted by propolis when first gathered by the bees, or when it is heated afterwards.

This substance is used to solder up all the cracks, flaws, and irregularities about the hive. A coat is spread over the inside throughout; when the hive is full, and many bees cluster outside the latter part of summer, they also spread a coating there. A layer seems to be annually applied, as old hives are coated with a thickness proportionate to their age, provided they have been occupied by a strong family.

It differs materially from wax, being more tenacious, and much harder, when old.

NEW SWARMS SOMETIMES USE WAX INSTEAD.

Our first swarms that issue in May or first of June, seldom use much of the pure article for soldering and plastering, but use instead a composition, most of which is wax. I have noticed at this season, when pieces of old hives were left in the sun, that this old propolis would become soft, and that the bees would gather it, packing it on their legs, the same as they do pollen. It is detached in small particles, and the process of packing can be distinctly seen, as the bee does not fly during the operation, as when packing pollen.

If you squeeze a piece of dry comb together, in a hot day, making a compact ball, and leave it near the hives, the Italians will carry it away in a short time; using it as they do propolis.

MORE ABUNDANT IN AUGUST.

In August they use a hundred-fold more propolis than in June, and at this time they manifest no disposition to gather any from the old boards, etc. It would seem that they prefer the new article which they now have in abundance. Boxes filled in June, contain but very little, sometimes none. But when filled in August, the corners, and sometimes the top and sides are well lined. Cracks, large enough for the bees to pass through, are sometimes completely filled with it.

The reason of its being collected in greater abundance at this season may perhaps be found in the fact, that the buds of trees and shrubs are now generally formed.

So few bees are engaged in collecting it, that it is difficult to detect one in the act, but by persevering watchfulness, they may sometimes be seen, particularly in August.

This is but one of the many branches of apiarian science, which are sadly neglected. So much error is mingled with the truth, that nothing but the most patient scrutiny, and untiring investigation, can separate them.

CHAPTER XIX.

TRANSFERRING.

If there is no other object in transferring bees intended to be kept in swarming hives, than simply to have them in a situation to look at, it is of doubtful utility. But when any other motives can be adduced, it will pay better. If you wish to Italianize your bees, or to get them in such condition that you can readily and safely make artificial swarms, you probably cannot do better than to change a part of them to movable comb-hives. It is not a very formidable operation to transfer combs, brood, and honey, as well as bees.

PREPARATION.

Prepare the frames without the triangular bar at the top, as directed in the chapter on hives. A part of them should have a second bar, one inch wide, by one-fourth inch thick, some placed a little above, and others a little below the centre. For each frame you will want from two to four pairs of thin narrow strips—one-fourth inch square will do. They should be just half an inch longer than the height of the frames, projecting one-quarter inch above and below. Cut a small notch in each end to hold a piece of twine, with which they are to be tied together.

Fig. 31.—PIECE USED TO STEADY THE COMBS IN TRANSPORTATION AND TRANSFERRING.

TIME.

April is usually the best season. There is then the least honey in the way, and not often much brood. Two or three weeks after the first swarm, is also a good time.

It is practicable at any time, except in very cold weather. The more honey and brood there are present, the more care is required.

HOW TO DO IT.

When ready to operate, if the weather is cool, warm a couple of empty hives, and choose, if possible, a strong colony with straight combs. Take the hive from the yard, and set one of the empty hives on the stand to catch any bees returning from the field, if you operate when any are flying, and proceed to drive out the bees as directed in the chapter on pruning. After the bees are out of the way, take the hive to a warm room and remove one side. Have ready a wide board, on which are spread smoothly several thicknesses of cloth, on which lay two or three of your small strips; and having loosened a comb from its fastening, lay it upon them. The yielding surface will prevent the bruising of the combs. Now measure the comb with the frame; if large enough to fill it, mark and trim it off exactly to fit. If too small, take one of the frames that has a partition through it; it will fit some of them. When the frame is filled, whether with one piece or several, tie enough of the strips on each side to hold the comb in place, and raise the frame to a perpendicular position. I have used twine to wind around the frame and combs, instead of fastening with wood, but do not like it as well. These pieces of wood are also just what is needed to steady the combs in full colonies, when they are to be transported long distances.

KEEP BROOD TOGETHER.

When transferring combs containing brood, it will not do to separate them,—scattering it all through the hive—unless in hot weather. In cool weather the bees must keep it warm, which they cannot do unless it is all in one

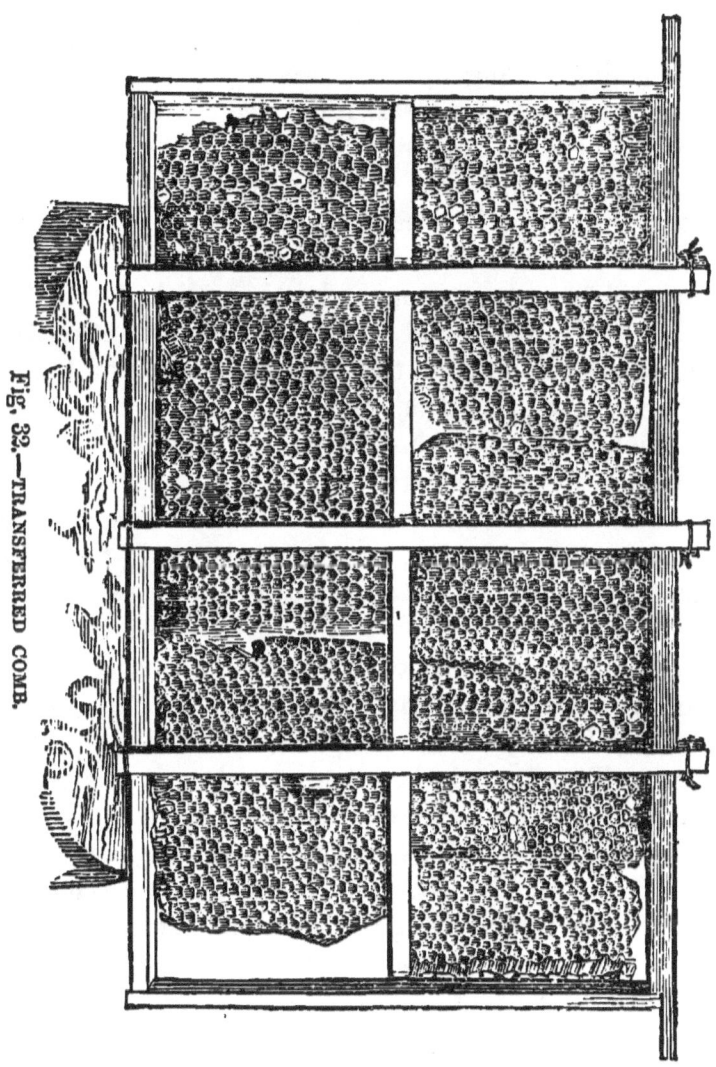

Fig. 32.—TRANSFERRED COMB.

place. Let the combs occupy the same relative position that they did in the original hive.

CAUTION.

When all is arranged, if no bees are flying, set the new hive on the floor in a dark room, with the front raised, shake the bees down by it and they will readily enter. The next morning, return it to the old stand. If bees are at work and there is a scarcity of honey, get what bees you have in the hive into which they were driven, to enter the new hive, and as soon as they have licked up all the dripping honey, which will take an hour or two, return them to the stand, taking care that no honey is left on the outside of the hive to attract robbers. A very little might excite plundering, and the transferred colony is in a bad condition to repel such attacks.

It would be well to almost close the entrance, allowing room for only one bee to pass at a time. If the colony is a strong one and covers all the combs, they will weld all the combs together wherever they touch each other, and also where they come in contact with the frames, in a very few days, when the frames may be taken out, and the splints that held the combs in place, removed.

There will frequently be occasions when it is desirable to transfer only the bees to a new hive, or to a hive with empty combs. The combs, where the bees are, may be old, or contain diseased brood. If it is only old, and is healthy, you need only to drive out the bees, and introduce immediately. But should it be otherwise, as when the bees have been wintered in diseased combs, with the intention of transferring in early spring to hives containing combs and honey, reserved for this purpose in the fall, it can be done thus. First, drive the bees into an empty hive previously made warm. If all do not go in readily, break out the combs and brush them in. A less number of bees will be wasted if you work in a dark room, and use a candle or lamp. If preferred, there would be no harm done, if the bees were first paralyzed with Puff-ball, yet the combs might have to be broken out to get the few bees sticking between them, at this season—spring—it is well to save every bee. Keep them in the empty hive until the honey taken with them from the diseased hive is consumed—thirty-six or forty-eight hours. The hive to receive the bees permanently, should be brought into a warm room several hours previously. If bees are to be put into an empty hive, it should be done in warm weather.

CHAPTER XX.

SAGACITY OF BEES.

On this subject I have but little to say, as I have failed to discover any individual manifestations of shrewdness or sagacity, that all swarms would not exhibit under similar circumstances.

TOO MARVELOUS.

Writers, with a great love of the marvelous, are not content with their astonishing displays of instinct, but must add a share of reason to their other faculties, and profess to discover "an adaptation of means to ends that reason alone can produce." It is very true, that without close inspection, and comparison of the conduct of different swarms in similar cases, one might arrive at such a conclusion. It is difficult, as all will admit, "to tell where instinct ends, and reason begins." Instances of sagacity like the following, have been recorded.

INSTANCES OF SAGACITY.

"When the weather is warm, and the heat inside is somewhat oppressive, a number of bees may be seen stationed around the entrance, vibrating their wings. Those inside will turn their heads towards the passage, while those outside will turn theirs the other way. A constant agitation of the air is thus created, thereby ventilating the hive more effectually."

All populous stocks do this in hot weather.

Again. "A snail had entered the hive, and fixed itself against the glass side. Being unable to penetrate it with their stings, the cunning economists fixed it immovably by merely cementing the edge of the orifice of the shell to the glass with resin (propolis), and thus it became a prisoner for life."

Now, the same instinct that prompts the collection of propolis in August to fill every crack, flaw, and inequality about the hive, would use it to cement the edges of the snail-shell to the glass, and a small stone, block of wood, or any substance that they are unable to remove, is fastened with it in the same manner. The lower edges of the hive, when in close proximity to the bottom, are joined to it with this substance. The stoppers of the holes in the top are thus fixed in their places; and the published instance of the unaccountable "sagacity," that once fastened close a little door on the top of a hive, may be nothing more than a result of this instinctive habit.

Other instances of "wonderful sagacity" will, I think, also be found simply to be individual examples of their common customs.

Whenever the combs in a hive have been broken, or when they have been added, as mentioned in Fall Management, the first act of the bees, is, to fasten them in their present position. When the edges are near the side of the hive, or two combs are in contact, a portion of wax is detached, and used for joining them together, or to the side. Where two combs do not touch, and yet are close together, a small bar is constructed from one to the other, preventing any nearer approach. This may be observed by tipping the hive a few inches from the perpendicular, after being filled with combs in warm weather, and permitting it to remain thus for a few days.

Should nearly all the combs in the hive become detached from any cause, and fall to the bottom in a mass of ruin, their first steps are, as just described, to construct pillars to keep them in the same shape. In warm weather they will, in a few days, have made passages throughout every part of the mass, by biting away the combs where they are in contact. Little columns of wax below, support the combs above, irregular to be sure, but as scientific as circumstances will permit. The whole will be firmly fastened together, and not a single piece can be removed without breaking it from the others.

NO PART OF THE HIVE INACCESSIBLE.

A piece of comb filled with honey, and sealed up, may be put in a glass-box with the ends of these sealed cells, touching the glass. The principle of allowing no part of their tenement to be inaccessible, is soon manifested. They immediately bite off the ends of the cells, remove the honey that is in the way, and make a passage next to the glass, leaving a few bars from it to the comb, to steady and keep it in its position. A single sheet of comb lying flat on the bottom-board of a populous swarm is cut away on the under side, by passages in every direction, numerous little pillars of wax being left for its support.

How any person in the habit of watching their proceedings, with any degree of attention, could conclude that the bees *raised* such comb by mechanical means, and then put the props under for its support, is somewhat singular.

These things are none the less wonderful, when considered as the result of instinct. I am not sure but the display of wisdom is even greater than if the power of planning their own operations had been given them.

I have mentioned these examples to show that a course of action induced by the peculiar situation of one family, would be adopted by another in a similar emergency. Were I engaged in a work of fiction, I might let fancy reign, and endeavor to amuse, but this is not my object.

WE SHOULD BE CONTENT WITH FACTS.

Let us endeavor to be content with truth, and not murmur that her marvels are no greater.

When we remember that the material for their combs is formed in the rings of their own bodies, and that, untaught, they detach it, and construct combs of the most beautiful symmetry, and that unbidden they go forth to the field to gather stores for the future use, we can but perceive that throughout the whole cycle of their operations, one law and one power governs, and whoever would seek that directing power, must look beyond the sensorium of the bee.

CHAPTER XXI.

SELECTING COLONIES FOR WINTER.

FIRST CARE.

When the flowers fail at the end of the season, it is necessary to ascertain which are the weakest stocks, and all that cannot defend themselves should either be removed or reinforced. The strength of all stocks is apt to be thoroughly tested within a few days after a failure of honey.

Should any be found with too few bees for defence, they are quite sure to be plundered. Hence the necessity of *immediate* action, that we may secure the contents in advance of the robbers.

STRONG COLONIES INCLINED TO ROB.

Strong stocks, that during a yield, have occupied every cell with brood and honey, will, when it fails, soon have cells left empty by the hatching of the young bees. The want of honey to fill these empty cells, appears to be a source of much uneasiness. Although such hives may be well stored, I have ever found them much more disposed to plunder, than weaker ones with but half the honey. As very feeble families cannot be strengthened now, it is best to remove them at once, and put temptation out of the way. Carelessness is but a sorry excuse for allowing bees to establish this habit of dishonesty. Should any stocks be weak from disease, the consequences would be even more disastrous than bad habits. The reasons why such impure honey should not go into thrifty stocks, have already been given. If we want the least possible trouble with our bees, none but the best should be selected for winter.

REQUISITES OF GOOD STOCKS.

But the requisites of a good stock, seem to be but partially understood. Judging from the number lost annually, too many bee-keepers are careless or ignorant in making the selection. They seem to think that because a stock has once been good, it will remain so. The condition of bees is so changeable, especially in the summer and swarming season, that this idea should never be entertained. *We must know their present condition by actual examination.*

The requisites of a good stock are, a hive of proper

shape and size, (viz.: 2000 cubic inches) full of combs, well stored with honey, a large family of bees, and freedom from disease, which must be ascertained by actual inspection. The age is not important, until they are ten years old. Stocks possessing these essentials can be wintered with but little trouble. But it cannot be expected that all will be in this condition. Many bee-keepers wish to increase their numbers and would like to keep all that they can, practicably. Many deficiencies can be supplied, with a little attention, and I shall endeavor to show that it is profitable to do so until the number of bees kept is too great for the supply of honey.

All can understand why it is a loss for bees to eat honey part of the winter and then die—that the honey consumed might have been saved, and that it makes no great difference to the bees, whether they are killed in the fall, or starve in the winter. I am not an advocate for fire and brimstone as the reward of all unfortunate stocks, and shall recommend it only when it is unavoidable. We will see how far it can be dispensed with.

DISADVANTAGE OF KILLING BEES.

Those rustic bee-keepers who are in the habit of making their hives large enough to hold from 100 to 150 pounds, killing the bees in the fall, and sending the honey to market, will probably continue the use of sulphur, unless we can convince them that it is far better to make the hive smaller, and have fifty or eighty pounds of this honey in boxes, which will sell for more money, and at the same time save their bees for stock hives. When the hives are of the proper size, the honey is not equivalent in value to the bees.

CAUSE OF POOR COLONIES VARIES IN DIFFERENT SECTIONS.

The particular deficiency of weak stocks depends somewhat on the section of country. Where the principal source of honey is Clover or Basswood, it will fail, partially at least, before the end of warm weather. Some poor or medium stocks will continue to rear brood too extensively for their means, and exhaust their winter stores in consequence; such will need a supply of honey.

But where great quantities of Buckwheat are sown, cool weather follows almost immediately after the yield, and stops the breeding. Consequently, in such localities, a scarcity of bees is more common than a lack of honey. If there are bees enough at the 1st of September, there will usually be plenty of honey. There are exceptional cases in all localities.

It is common to have stocks with stores amply sufficient to carry a good family through the winter, and too few bees to last till January, or even to defend themselves from robbers, hence bees are more frequently to be supplied, than honey.

POOR STOCKS MAY BE UNITED.

It is usual to have a few hives with too little honey, as well as too few bees. It is very plain, that if the bees of one or more of this class were united successfully with some of the former, we should have a respectable family. I have thus united stocks that proved first rate.

WHEN IT IS NOT BEST.

Whenever we make additions in this manner, it would be well first to ascertain the cause of a scarcity of bees; if it is over-swarming, barren or drone queen, or *loss* of queen, it is well enough; but if it is from disease, reject them, unless the bees are to be transferred the next spring, and even then, if so many cells are occupied with dead brood, that the bees cannot be successfully wintered. The greatest difficulty in uniting two or more families in this manner, arises from their belonging in the same apiary, where they have marked the locations. It has been sufficiently shown that bees will return to the old stand.

To prevent bad results, it has been recommended to " set an empty hive with some pieces of comb fastened in the top, in the place of the one removed, to catch the bees that go back to the old stand, and remove them at night, a few times, when they will remain." This should be done only when we cannot do better; it is considerable trouble, and is not always satisfactory.

I like to bring them a mile or more for the purpose, and thus avoid all such trouble. Two neighbors that distance apart, each having stocks in this condition might exchange bees, with mutual benefit. I have done so, and considered myself well paid. But, latterly, I have several apiaries away from home, and have no difficulty.

The practice of making one good stock out of two poor ones cannot be too highly recommended. Aside from its advantages, it relieves us from all disagreeable feelings in regard to taking life.

TWO SWARMS UNITED, EAT LESS THAN WHEN SEPARATE.

Even when a stock already contains bees enough to insure it for winter, another of the same number of bees may be added, and they will not consume five pounds more than one swarm would if kept alone. If they should be wintered in the cold, the difference might not be one pound. Why a larger number of bees does not consume a proportionate quantity of honey, (which the experience of others, as well as my own, has thoroughly proved,) is a mystery. If the fact that a greater number of bees generates more animal heat, and they therefore eat less, is a solution, it is a powerful argument for keeping bees warm in winter.

Notwithstanding this, a *good* stock is not made any better by doubling the number of bees. I have tried the experiment thoroughly, and when they commenced work the next spring, such double families promised much; but when the swarming season arrived, good single swarms were in the best condition. I am unable to give any satisfactory reason for this. Stocks, which have cast no swarms, are no better the following spring than others. The same cause may operate in both cases. It therefore appears unnecessary to unite two or more good swarms, unless we are particularly sensitive about killing the bees.

SEASON TO OPERATE.

The usual season for operating is, when all brood has matured and left the cells. The exceptions are where there are not bees enough to protect the stores, in which cases it may be necessary, immediately after the failure of honey in the flowers.

PARALYZING THE BEE.

Col. H. K. Oliver, of Mass., has the credit of inventing the fumigator, an instrument to burn Puff-ball. By the aid of this, the smoke is blown into the hive, paralyzing the bees in a few minutes. They fall to the bottom apparently dead, but will recover in a few minutes, on receiving fresh air.

DESCRIPTION OF FUMIGATOR.

The fumigator is made of a tin tube, four inches long and two in diameter, with a stopper of soft wood, three inches long, exactly fitting one end when driven in half an inch, and secured by little nails driven through the tin. Through the centre of this stopper there is a hole one-fourth inch in diameter. To prevent this hole from filling up, the end in the tube is covered with wire-cloth, bent a little convex. The end of this stopper is cut down to about half an inch, tapering from the tin. For the other end a similar piece of wood is fitted, though a little longer, and not to be fastened, as it must be taken out at every operation. The outer end of this is cut down to a shape to be taken into the mouth, or attached to the pipe of a pair of bellows. It can be made wholly of tin, but then it is necessary to use solder, which is liable to melt and cause leaks. The puff-balls must not be too much injured by re-

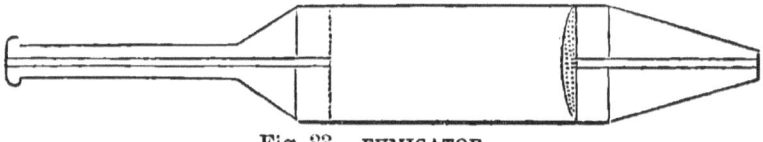

Fig. 33.—FUMIGATOR.

maining in the weather, and should be taken, if possible, just before they are ripe and burst open. When not thoroughly dry, put them in a warm oven. Remove the rind carefully, ignite it with a live coal—it will not blaze—blow it and get it thoroughly to burning before putting it in the tube. Put in the stopper and blow through it; if it smokes well, you are ready to proceed. When it does not burn freely, unstop and shake it out.

HOW TO OPERATE.

Invert the hive to receive the bees, set the other over it, and stop all crevices to prevent the escape of the smoke. Insert the end of the fumigator into a hole in the side of the hive, blow at the other end, and in two minutes you may hear the bees begin to fall. Both hives should be smoked, the upper one the most. The other only needs smoke enough to make the bees similar in scent to those introduced. At the end of eight or ten minutes, the upper hive may be raised, and any bees sticking between the combs, brushed down with a quill. The two queens are of course together; one will be destroyed, and no difficulty arise. But if either of them is young, and you have been convinced that such are more prolific, and happen to know which hive contains her, you can preserve her by a little variation of the process.

Instead of inverting one hive, set them both on a cloth right side up, and smoke the bees, and while they are paralyzed, search out and preserve the desired queen. Then put the bees all together. Tie a thin cloth over the bottom of the hive to prevent the escape of the bees. Before they are fully recovered, they seem rather bewildered, and some will get away. Set the hive right end up, and raise it an inch; the bees drop on the cloth, and fresh air passing under soon revives them. In from twelve to twenty-four hours they may be let out.*

Families united in this way will seldom quarrel, but will remain together, defending themseves against intruders as one swarm. I once had a stock nearly destitute of bees, with abundant stores for wintering a large family. I had let it down on the floor-board, and was on the lookout for an attack. The other bees soon discovered their weakness and commenced carrying off the honey. I had brought home a swarm to reinforce them only the day before, and immediately united them by means of the fumigator. The next morning I let them out, allowing them to issue only at the hole in the side of the hive. It was amusing to witness the apparent consternation of the robbers that were on hand for more plunder; they had been there only the day before, and had been permitted to enter and depart unquestioned. But, lo! a change had transpired. Instead of finding open doors, and a free passage, the first bee that touched the hive was seized, and very rudely handled, and at last, dispatched with a sting. A few others receiving similar treatment, they began to exercise a

---

* When a condemned colony is smothered in the old hive, so many of the bees remain sticking among the combs, that it is very unpleasant work to remove the combs. It is better on the whole to drive out the bees in the beginning, and the few remaining on the combs can be paralyzed, and will drop down without difficulty.

little caution, and tried to find admittance on the back side, and other places. They even attempted to enter one or two other hives, on either side, perhaps thinking they were mistaken in the hive, but these being strong, repulsed them, and they finally gave it up.

HOW BEES WERE WINTERED AFTER A SCARCITY OF HONEY.

One season, some years ago, when I was anxious to increase the number of my stocks to the utmost, I had none but early swarms, with sufficient honey for winter. Twenty-five pounds is required for winter stores in this section. I had over thirty young swarms with less than that quantity.

There was a large number of old colonies with diseased brood, and but *few* bees, but sufficient honey. Such honey is not deleterious to the old bees, hence I transferred the bees of these new swarms, by paralyzing, to the old stocks containing black comb and diseased brood. The bees were thus wintered on honey of but little account, and all the honey that was in the healthy hives was saved. These new hives were set in a cold dry place during winter, *right end up*, to prevent the honey from dripping out of the cells. Some will leak then, but not as much as when the hive is bottom up. Honey that runs out, when the hive is bottom up, will soak into the wood at the base of the combs, and has a tendency to loosen them, and render them liable to fall, etc. The bees should be returned to the new hives the following March. For the method of transferring see Chap. xlix.

ADVANTAGE OF TRANSFERRING.

When a good-sized family is put in a hive containing fifteen or twenty pounds of honey, and nearly half full of clean new comb, they are about as sure to fill up and cast a swarm, as one that is full, and has wintered a colony.

One cause of superior thrift may be found in the circumstance that all moth eggs and worms are frozen to death, and the bees are not troubled with any worms before June. No young bees have to be removed to work them out, hence nearly every young bee that is fed and sealed up, comes forth perfect, which makes a vast difference in the increase. Any person wishing to increase the number of his colonies to the utmost, will find this plan of saving all partly filled hives, much more profitable than to break out the honey for sale. Suppose you have an old stock that needs pruning, and you have neglected it, or it has refused to swarm, and you have had no opportunity to do it, without destroying too much brood. You can let it be, and put on the boxes—perhaps get twenty-five pounds of cap honey—and then winter the bees in it as described, and transfer them to the new combs in the spring.

Again, if there are no old or diseased stocks to be transferred in the spring, keep them until the swarming season. If a swarm put into an empty hive would just fill it, the same swarm put into one containing fifteen pounds of honey, would evidently make that quantity of surplus. The advantage is in the comparative value of box and hive honey; the former being worth from thirty to a hundred per cent. more.

UNITING COMB, HONEY, AND BEES.

I have occasionally adopted another method of making a good stock from two poor ones, which the reader may prefer. When all the old stocks have been reinforced that need it, and there are still some swarms with too few bees and too little honey to winter safely in their present condition, two or more can be united. The fact, which has been sufficiently tested, that two families of bees, when united and wintered in one hive, will consume but little, if any more, than each would separately, has an important bearing upon this case. If each family should have fifteen pounds of honey, they would consume it all, and probably starve at last—making a loss of thirty pounds. But if the contents of both were in one hive, there would be an abundance, with some to spare in the spring.

The process of uniting them is simple. Smoke both of the hives, and then turn them over. Choose the one with the straightest combs, or the one nearest full, to receive the contents of the other; trim off the ends of the combs square across, and it is ready. Remove the sticks from the other, and with your tools, take out the combs with the bees on as before directed, one at a time, and carefully set them on the edges of the other combs. If the shape will admit, let the edges match, if not set them crosswise. Bits of wood or rolls of paper should be put between, to preserve the right distance. When both hives are of one size, the transferred combs will exactly fit, if you are careful to adjust them properly. You will now want to know "what is to prevent these combs from falling out when the hive is turned over?" It is not to be turned over, but is to remain bottom up in some dark place for some time, or until spring. (See Wintering Bees.) The bees will immediately join the combs together. The hive being inverted, the honey in these combs will be consumed first, and when the hive is set out in spring, it will be a rare occurrence for any pieces to drop out. Should any combs project beyond the bottom of the hive, they may be trimmed off even, after they are fastened, before they are set out. An additional cross-stick may pass under the bottom of the combs, to support them if you desire. You will probably never discover any difference in their subsequent prosperity in consequence of the crossing of the combs in the middle. I have had them thus, that were some of my most prosperous colonies.

As this operation need not be performed until November, there is still another advantage; families of the same apiary can be united, and will mostly forget the old location by spring, hence no difficulty will arise when they are returned to their old stand.

In some sections, honey is more frequently wanting than bees or comb; in such cases it is an advantage to feed, until enough is stored for winter. This can be determined by what the hive actually weighs when done. It is insufficient to simply weigh the honey that is fed.

WHEN IT IS BEST TO FEED.

It should be done in October, that the honey may be sealed up before cold weather. If done before, too many of the combs will be occupied with brood. Feed as fast as possible, that they may not start too much brood.

But if they lack comb as well as honey, and you still wish to try feeding, it should be done, if possible, in warm weather, as they cannot make combs advantageously in the cold.

Directions for feeding will be found in CHAP. VII.

If your hives are not full, and are to be wintered in the house, bottom up, they may be fed at any time during the winter by merely laying pieces of comb containing honey, on those in the hive. The bees readily remove the contents into their own combs. When empty, replace them, until they have a full supply. They will join such pieces of comb to their own, but there will be no harm in breaking them loose. The principal objection to feeding in this way, will be found in the tendency to make them uneasy and disposed to leave the hive, when it is desirable to have them as quiet as possible. A thin cloth will be necessary to confine them to the hive.

With the movable combs, a great deal of the foregoing trouble may be avoided. A large proportion of the colonies will almost always store more honey than needed for winter, to an absolute disadvantage. The Italians are quite sure to have an excess, in ordinary seasons. Seventy pounds was quite a common surplus, in the summer of '64.

There will be some late or small swarms, or some that were divided too late to obtain sufficient winter stores. By taking a comb or two from such heavy hives, and exchanging with the light ones, all are benefitted. The light hives are made fit to winter, and the others are better off because the bees can have room in the empty cells, to pack themselves closely for mutual warmth in severe weather.

When a season occurs like the summer of '63 in this section, when but few colonies have enough stores, a great many only half enough, and some, still less,—unless it is decided to feed all—some of the lightest may be taken up, and the heaviest combs appropriated for the use of those to be wintered. By attention through the winter and spring, occasionally giving a comb containing three or four pounds for an empty one, as they need, they can be wintered without any great deal of trouble.

I have now given directions to avoid killing any family of bees worth saving. When such as need feeding have been fed, and all weak colonies have been made strong by additions, etc., but little more fall work is needed in the apiary.

It is only when there are weak stocks unfit for winter in any shape, that it is necessary to be on the look out every warm day to prevent pillage.

CHAPTER XXII.

STRAINING HONEY AND WAX.

REMOVING COMBS.

The combs of the movable frame hive are so easily lifted out, and removed from the frames, and the bees shaken off, that it is quite unnecessary to give explicit directions for removing the honey from such, for straining.

The most convenient method of removing combs from the box hive is by taking off one of its sides, but if it is properly nailed, this is apt to split the boards, and injure it for subsequent use. With tools such as have been described in Chap. xiii, it may be done very nicely without injury to the hive. The chisel should have the bevel on one side, like those used by carpenters. The flat side is placed next the board of the hive and the bevel, crowded by the combs, will follow it close the whole length; with the other tool the combs are cut across the top, and readily lifted out. If preferred, they may be cut across near the centre, and half a sheet taken out at a time; this is sometimes necessary on account of the cross-sticks.

Such combs as are taken from the middle, or the vicinity of brood-cells are generally unfit for the table, and should be strained.

METHOD OF STRAINING.

There are several methods of performing this operation. One is, to mash the comb and put it in a bag, and hang it over some vessel to catch the honey as it drains out. This will do very well for small quantities in warm weather, or in the fall before any of it is candied. Another method is to put such combs in a colander, and set this over a pan in a moderately hot oven. This will melt the combs, and the honey and a portion of the wax will run out together. The wax will rise to the top and cool in a cake. It is liable to burn and require care. Many prefer this process, as there is less taste of bee-bread, no cells containing it being disturbed; but unless it is stirred, some of the honey will not drain out. If desired, two qualities may be made. Another way is to break the combs well in a colander and allow the honey to drain out without much heat, and afterwards skim the small particles of comb that rise, or pass the honey through a cloth or piece of lace.

But for large quantities, it is well to have a special arrangement. Make a box four inches high, and sixteen or eighteen inches wide, by four feet in length, with a wire-cloth bottom. Put legs under, to raise it to a convenient height. Under the strainer, place a board same in width and length, elevated at one end, to catch the honey and conduct it into some vessel. Break up the comb with the hands, but do not stir it unnecessarily, or the bee-bread will be mixed with the honey. The box may be filled with this, and will hold one hundred pounds or more. The honey as it first runs through, will contain particles of comb, but if left to stand for a day or two, these will all rise to the surface. Through a hole near the bottom, the pure honey may be drawn off.

When you have obtained all that will readily drain out, put the comb in a barrel or large box with a few holes in the bottom. A large quantity will warm up of itself, and another portion of honey will run out, of somewhat inferior quality.

Should the weather be somewhat cool, it will become thickened too quickly to run out well, when a press of some sort is of great advantage. Unless the weather is warm, the bees should be removed only on the day the comb is broken up.

METHEGLIN AND VINEGAR.

When no more honey can be conveniently obtained, a still further saving may be made, by covering the comb with boiling water, stirring well and draining; when the comb should be made into wax at once to prevent moulding.

The sweetened water should be boiled and skimmed till clear. If strong enough to raise a potato one-third above the surface, it may be put away for metheglin, which will be fit for use in twelve months, and improve with age. By reducing it considerably, vinegar may be made in the same manner as from other material.

FEEDING REFUSE HONEY.

There is still another method of saving the honey remaining in the comb after draining. Take a large movable comb-hive—it would pay to make one on purpose with a large number of frames—full of empty combs or frames. Make a second hive without top or frames; cover the bottom an inch thick with the refuse comb, and set the first hive containing bees and combs over it. As the bees have to work in comparatively cool weather, it is well to make a strong swarm by uniting three or four condemned colonies. They should not be allowed any queen to fill the combs with brood, but may have a little brood to content them with the idea of raising one. In a few hours, they will lick up several pounds, leaving the comb perfectly clean, when more may be added, and this process continued till all the comb is clean. By this time they may have fifty or one hundred pounds of beautiful honey.

If this honey is from hives that contained diseased brood it would be best to secure it from the other bees; but if otherwise it would answer every purpose for wintering bees, and this swarm might remain. In this case, it would be best to let them have a queen, by all means, in order to have them build worker combs.

In warm weather, if a door or window leading to a room where honey is exposed, is left open, the bees will find their way in, during a scarcity of honey. Doors and windows should be kept closed, allowing them no means of entrance.

MAKING WAX.

Several methods have been adopted for separating the wax. I have never been able to secure the whole. Some recommend heating it in an oven, similarly to the method of straining honey through the colander, but I have found it to waste more than when melted with water. A better way for small quantities, is to fill a coarse stout bag, half full of refuse comb, with something to sink it, and boil it in a kettle of water, pressing and turning it frequently, till the wax ceases to rise.

QUANTITY WASTED.

When the bag is emptied, by squeezing a handful of the contents, the particles of wax may be seen, and you may judge of the quantity wasted.

LARGE QUANTITIES.

For large quantities the foregoing process is rather tedious. It can be facilitated by using two levers four or five feet long, and about four inches wide, fastened together at the lower end by a strong hinge. The comb is put into a kettle of boiling water, and will melt almost immediately; it is then put into the bag, and taken between the levers, in some large vessel, and pressed. The contents of the bag should be shaken and turned several times during the process, and if need be, returned to the boiling water, and then squeezed again. The wax with a little water is now to be re-melted, and strained again through finer cloth, into vessels that will mold it into the desired shape. As the sediment settles to the bottom of the wax, when melted, a portion may be dipped off nearly pure without straining.

By adding an acid to the water in which the wax is melted, it may be separated much more readily. A quart of vinegar to a gallon of water, or a small spoonful of nitric acid is sufficient. There may be a little less labor, where it is available, in using a heavy press like one made for cheese, with a hoop made of slats an inch wide and one-fourth inch apart, firmly bound. The bag with melted comb is put into this, and throughly pressed, while boiling hot.

Wax may be bleached in the sun, in a short time, in cool weather, but it must be in very thin flakes. It is readily obtained in this shape by dipping a thin board or shingle, thoroughly wet, into pure melted wax; enough will adhere to make it the desired thickness, and will cool instantly on being withdrawn. Draw a knife along the edges, and it will readily cleave off. Exposed to the sun in a window or on the snow, it will become perfectly white, when it can be made into cakes for market, where it commands a much higher price than the yellow.

I presume there are chemical processes by which this result is obtained, but I am not familiar with them.

CHAPTER XXIII.

WINTERING BEES.

There is almost as much irreconcilable diversity of opinion with respect to wintering bees as in the construction of hives.

DIFFERENT METHODS.

We are told to keep them warm, and to keep them cold; to keep them in the sun and out of the sun; to bury them in the ground; to put them in the cellar, in the chamber, in the wood-house, and to do nothing with them. Here are plans enough to drive the inexperienced into despair. Yet I have no doubt but bees have occasionally been successfully wintered by all these contradictory methods. That some are superior to others, needs no argument; but which are *best*, is our province to inquire. Let us endeavor to investigate the subject without prejudice.

We will first examine the condition of a stock left to nature, without any care, and see if it affords us any hints as to how we shall protect them. By close observation we shall probably discover that the oft-repeated assertion, that bees will never freeze except when without honey, has led to many errors in practice.

WARMTH FIRST REQUISITE.

Warmth being the first requisite, a family of bees crowds together at the approach of cold weather, into a compass corresponding to the degree of cold. Those on the outside are somewhat stiffened with cold, while those within, are as brisk and lively as in summer. In severe weather, every bit of space within their circle is occupied;

even each cell not containing honey or pollen holds a bee. Suppose this cluster is sufficiently compact for mutual warmth, with the mercury at 40°, and it falls to zero in a few hours, this body of bees, like most other things, will speedily contract. Some bees on the outside, being already chilled, do not keep up with the shrinking mass, and being left exposed at a distance from their fellows, receive but little benefit from the warmth of the cluster, and are soon frozen.

SIZE OF COLONY.

A good colony will form a ball or sphere about eight inches in diameter, generally about equal every way, which must occupy the spaces between four or five combs. As the combs separate them into divisions, the two outer ones are the smallest and most exposed; these are often found frozen to death in severe weather. Should evidence be wanting to show that bees will freeze, the above seems to furnish it.

Suppose a quart of bees to be put in a hive where all the cells are lengthened out, and filled with honey, there would be only one-fourth inch spaces between the combs, leaving room for only one course of bees. The combs are perhaps one and a half or two inches thick. These single layers of bees could not of course keep themselves properly warm. Although every bee would have food in abundance without changing its position, the first severe weather would probably destroy them all. But I admit this to be a rather improbable condition.

PROMOTION OF WARMTH.

Their winter quarters are among the brood combs, where the hatching of the brood leaves most of the cells empty, and there is a half inch space between the combs; a wise and beautiful arrangement, as ten times as many bees can pack themselves within a circle of six inches, as in the other case, and in consequence, the same number of bees can secure much more animal heat and endure the cold much better. But a *small* family, even here, will often be found frozen, as well as starving.

Besides freezing, there are some other facts to be observed in connection with stocks that stand in the cold.

MOISTURE.

Physiologists tell us "that innumerable pores in the cuticle of the human body, are continually throwing off waste or worn-out matter; that every exhalation of air carries with it a portion of water from the system, unperceived in warm weather, but condensed into particles large enough to be seen in a cold atmosphere." Now, if analogy be allowed here, we will say that the bee throws off waste matter and water in the same way. Its food being liquid, nearly all will be exhaled; this passes off in moderate weather, but in cold it is condensed, and the particles lodge on the comb and sides of the hive in the form of frost, which accumulates as long as the weather is very severe. In the middle of the day, or as soon as the temperature is slightly raised, this begins to melt, first, near the bees, then at the sides of the hive. A succession of cold nights will prevent the evaporation of this moisture, and this process of freezing and thawing, will at the end of a week or two, form icicles as large as a man's finger upon the combs and interior of the hive. When the bottom is close to the floor, it forms a perfectly air tight sealing around the edges, and the colony is smothered. I have frequently heard bee-keepers say· "The snow blew in, and formed ice all round the bottom, and my bees froze to death." Others who have had their bees in a cold room, and found them thus, "could not see how the water could get there, any way—were quite sure it was not there when carried in, etc."

When the hive contains a very large or a very small family, there will be less frost on the combs; in the first case the animal heat will drive it off, and in the latter there will be but little moisture exhaled to freeze.

CAUSES OF STARVATION.

This frost frequently causes medium sized families to starve in cold weather, even when there is plenty of honey in the hive. If all the honey in the immediate vicinity of the bees is exhausted, and the combs in every direction are covered with frost, a bee leaving the cluster, and going among the combs for a supply would meet a fate as certain as starvation. And without timely intervention of warmer weather, all would perish.

Should bees escape starvation, there is another exigency often attending them in continued cold weather. I have said that small families exhaled but little moisture. Let us see if we can explain the effect in this case.

DYSENTERY.

In small stocks not enough heat is generated to exhale the aqueous portions of the food. The philosophy that explains why a man in warm blood, and profuse perspiration, exhales more moisture than when in a quiet state, will illustrate this. The bees under these circumstances must retain the water as well as the excrementitious part, which soon distends their bodies to an unendurable extent. Their cleanly habits, that ordinarily protect the combs from being soiled, are not a sure protection now, and they are compelled to leave the mass, very often in the severest weather, to expel this unnatural accumulation of fœces.

In a moderately warm day more bees will issue from a hive in this condition than from others. It would seem that part of them are unable to discharge their burden, their waight prevents their flying, and they get down and are lost. With the indications attendant upon such losses, my own observation has made me somewhat familiar, as the following will illustrate.

A neighbor wishing to purchase some stock hives in the fall, requested my assistance in selecting them. We applied to a perfect stranger whose bees had passed the previous winter in the open air. I found on looking among them, that he had lost some from the cause just mentioned, as the excrement was yet about the entrance of one old weather-beaten hive that was now occupied by a young swarm, and was about half filled with combs.

I saw at once what had been the matter, and felt quite confident that I could give its owner a correct history of it. "Sir," said I, "you were unfortunate with the bees that were in this hive last winter; I think I can give you some particulars respecting it."

"Ah, what makes you think so? I would like to hear your ideas; I will admit that there has been something peculiar about it."

"One year ago, you considered it a good stock-hive, it was well filled with honey, had a good family of bees, and was at least two or three years old. You had confidence that it would winter as well as any, but during the cold weather, somehow, the bees unaccountably disappeared, leaving but a very few, and they were frozen to death. You discovered it towards spring, on a warm day. When you removed the combs, you probably noticed a great many spots of excrement on them, as well as on the sides of the hive, particularly near the entrance. Also one-half or more of the breeding-cells contained dead brood in a putrid state, and this summer you have used the old hive for a new swarm."

"You are right, sir, in every particular. Now, I would like to know what gave you the idea that I had lost the bees in that hive. I can see nothing peculiar about that old hive, more than this one;" pointing to another that also contained a new swarm. "You will greatly oblige me if you will point out the signs particularly."

I then directed his attention to the entrance in the side of the hive; here the bees had discharged their excrement on the moment of issuing, until it was near an eighth of an inch thick, and two or three inches broad. It yet remained, and had just begun to cleave off. "You see this brown substance around the hole in the hive?"

"Yes, it is bee-glue, (*propolis*,) it is very common on old hives."

"I think not; if you will examine it closely, you will perceive that it is not so hard and bright; it already begins to crumble, and bee-glue is not affected by the weather for years."

"Just so, but what is it, and what has that to do with your guess-work?"

"It is the excrement of the bees. In consequence of many cells containing dead brood, the bees could not enter them, and they were unable to pack themselves closely enough to secure the animal heat requisite to drive off the water in their food, and it was therefore retained in their bodies, till they were distended beyond endurance. They were unable to wait for a warm day, and necessity compelled them to issue daily during the coldest weather, discharging their fæces the moment of passing the entrance, and sometimes, before, upon the combs. They were immediately chilled and could not return; the quantity left about the entrance shows that a great many must have come out. That they came out in cold weather is proved by its being left on the hive, because in warm weather they *leave* the hive for this purpose. The cluster inside was in this way so reduced that they were unable to keep from freezing."

"This is a new idea; at present it seems very probable;

I will think of it. But how did you know that it was an old hive, and that it was well filled?"

"When looking under it just now, I saw that combs of a dark color had been attached to the sides near the bottom, below where those are at present; this indicates that it had been full; and the dark color shows that the combs were old. Also, a swarm early and large enough to fill such a hive the first season, would not be very likely to be affected by the cold in this way."

"Why not? I think this hive was crowded with bees as much as any of my new swarms."

"I have no doubt that it appeared so; but we are very liable to be deceived by dead brood in the combs. A medium-sized family will make more show in such a hive, than a larger one that have empty cells to creep into, and can pack more closely."

"But how did you know about the dead brood?"

"Because old stocks are often thus reduced and lost."

"What were the indications of its having been filled with honey?"

"Combs are seldom attached to the side of the hive farther down than they are filled with honey. In this hive the combs had extended to the bottom, consequently must have been full; also, unless a colony is very much reduced, the hive is generally well stored, even when diseased."

"Why did you suppose it was near spring before I discovered it?"

"I guessed at that. The majority of bee-keepers are rather careless, you know, and when they have arranged their bees for winter, seldom give them much more attention, till they begin to fly out in spring."

"But what should I have done had I discovered the bees coming out?"

"As it was affected with dead brood, you could have

done but little; you would have lost it eventually. But had it been a colony otherwise healthy, and was thus affected only because it was small, or by the severity of the weather, you could have taken it to a warm room, and turned it bottom up, or given it abundant upward ventilation, and the heat would have converted most of the water contained in their food, into vapor. This would rise from the hive, and the bees could retain the excrementitious portion without difficulty till spring."

"I suppose you must get along without losing many through the winter, if I may judge from your confident explanations."

"I can assure you I have but little fear on this point. If I can have the privilege of selecting suitable stocks, I will engage to not lose one in a hundred."

"How do you manage? I would be glad to obtain a method with which I could feel as perfectly safe as you appear to."

"The first requisite is to have none but good hives. I unite weak families until they become strong, or make some other disposition of them." I then gave him an outline of my usual method of housing bees, which I can confidently recommend to the reader.

This accumulation of fœces is considered by many writers as a disease—a kind of dysentery. It is described as affecting them towards spring, and several remedies are given. If what I have been describing is not the dysentery, I have never had a case of it; but I think it is the same, and that inattention must be the reason that many do not discover it in cold weather, at the time that it occurs. Some stocks may be badly affected, yet not entirely lost, and moderate weather may arrest its progress. When a remedy is applied in spring, long after the cause ceases to operate, it would be singular if it were not effectual. I have no doubt but some have taken the natural discharge of fæces that always takes place in spring, when the bees first leave the hive, for a disease. Others, looking for a cause for diseased brood, and finding the hive and combs somewhat besmeared, have assigned this as sufficient; but according to my view, have reversed it, giving the effect for the cause.

There is some reason to suppose that moisture on the combs gradually mixes with the honey, making it thin, and that the bees will be affected as described, by eating so much water with their food. But some experiments have induced me to assign cold as an additional cause, as I have always found, when I put the hives where it was sufficiently warm, that an immediate cure was the result, or, at least, it enabled the bees to retain the excrement till set out in the spring.

WATER.

Much has been said recently, about furnishing bees with water during the winter, but the reasons for feeding it, and the results are so conflicting, that we have but little reliable evidence on the subject. My experience does not show that it is very efficacious when given as a preventive of dysentery, or necessary in rearing brood while housed.

After taking so much trouble to get rid of moisture, I am not disposed to recommend giving any more of it. I may be prejudiced and not qualified to judge, in consequence. It is also strongly urged that it is required when bees are rearing brood, and that they will speedily perish when deprived of it, from being shut up in the house—especially if they have candied honey.* I cannot even

---

* Mr. Harbison says: "I have had bees confined for a period of forty-eight days, about one-third of which time they were in a warm latitude, in transit to California; not a single drop of water did they get during all that time, and yet they reared and matured brood on the way, and it was found in some strong

understand the force or pertinency of the reasoning. How candied honey can make any difference when the bees do not eat it at all, I am unable to perceive. When a portion of honey is candied in the cells, the bees eat only that which is liquid, rejecting the rest; would they do more than take the liquid portion, were it all in that state? I doubt if there any disadvantages incident to the use of candied honey further than the waste. When the weather is warm enough, or the bees have increased enough to generate sufficient heat, it all liquifies—the bees eat it, and no harm arises. I think it quite likely that a very dry atmosphere would be detrimental to a colony of bees, as it is unnatural. It is probably best to avoid all extremes.

colonies, in all stages from the egg to those just emerging from the cells, on their arrival at Sacramento."

Mr. Harbison publishes a letter in his work, applying directly to this subject.

St. JOHNSVILLE, N. Y., *January* 4, 1860.

MR. HARBISON, *Dear Sir.*—In regard to the necessity of giving bees water during winter, I cannot say at present that my views are in accordance with those set forth by Mr. Langstroth on pages 342, 343, and 346 of his last edition. I fear that his remarks, and the translation from the German, by Mr. Wagner, will give very many inexperienced bee-keepers much unnecessary trouble. A constant supervision is indicated as necessary to safely take the bees through the winter. I do not remember as any plan was given to keep up a supply without attention. As a dearth of water is represented as the cause of much loss, of course those who take this theory for fact and expect success, must have some trouble to provide for these wants.

Not dreaming that water was essential to the health of bees in winter, I have for the last twenty-five years used my utmost endeavors to get rid of *all* moisture about the hive, and I have succeeded as effectually as any one. When put in the house, I open the holes in the top of the hive, and then invert it on sticks; a constant circulation of air through the hive carries with it *all* the moisture generated, the combs remain perfectly dry, and as far as I can discover, the bees are perfectly healthy. Instead of meeting a *general* loss with this method, I have wintered hundreds of stocks with a loss of less than two per cent. Why others, who take no pains comparatively, to ventilate, should suffer so much more loss than I do, I cannot comprehend;—that is, with this theory.

Many years ago, I became *fully* satisfied that the loss of nine-tenths of all the good colonies in winter, was a direct consequence of confining this moisture to the hive. The experience of every subsequent year gives additional evidence in favor of this idea.

Respecting the particles of candied honey found on the bottom-board, as in-

NATIVES OF WARM CLIMATE.

Bees being natives of a warm climate, need some assistance in maintaining a healthy condition throughout the winter. Let us see if we cannot keep them warm, save the bees, economize honey, and at the same time get rid of the excess of moisture. A large family expels it much better than a small one.

WARM ROOM.

When a large number of colonies is put together in a close room, the animal heat, from all combined, is an advantage—to the weak ones, at least. Yet the moisture is condensed in large drops, and can be seen on the sides of a glass hive. This excess of moisture is quite certain to mold the combs, and must be disposed of. Ample vents can be opened on the top, or the hive inverted. Any one can see that while the vapor is warm from the bees, it will all pass off, or so much of it, that the combs

dicating suffering for water, mentioned by Mr. Langstroth, I have been unable to arrive at a similar conclusion, because whenever the room in which they were wintered, was cold enough to candy the honey, I have invariably found the greater part of it, after the bees were set out and when they had abundant opportunity to get water. These particles may be seen at any time during spring, when the bees do not obtain sufficient honey from the flowers, for themselves and brood, and are necessitated to draw on their old stores. This seems very plain without the theory of need of water as may be readily seen. In each cell, only a part of the honey candies; the bees can swallow only the liquid portion, and must reject the other; this may be the case, although they fly out daily.

When the temperature of the hive becomes sufficiently warm to liquify this, it is no longer to be found.

I rather suspect that Mr. Langstroth has depended very much on the testimony of others in this matter of wintering bees. In his first edition of the "Hive and Honey Bee" in 1853, he recommended what he called a "protector" as *very* important. In his second edition he abandoned that plan, as not likely to pay, and suggested "special depositories." To show the advantages of this method, he quoted Dzierzon, and several pages from me explaining the manner of getting rid of this water. And now two or three years later, he supposes water is absolutely essential. In all our rural affairs there is no branch where there are more conflicting theories than in bee-culture, especially wintering them. No one can be *sure*, till he makes a few experiments of his own.

Your's Truly,

M. QUINBY.

will not mold. At the same time enough is diffused throughout the atmosphere of the room, to answer all the requirements of the bees. Twenty years experience has proved this principle correct.

One hundred colonies in any kind of wooden hives, cannot be wintered more safely and economically, than in a warm room. What I mean by economically, is, with the least consumption of honey. The room should be large enough to conveniently contain the number of hives to be wintered.

A small one in the dwelling house, or some out-building, or a warm dry cellar may be used. I have wintered them in all these places, and prefer the latter.

CELLAR PREFERRED.

If I were now to construct a room for this particular purpose, and could have just what I wanted, it would not differ materially from one I now use. It is a cellar under a barn on a side hill, where the ground descends just enough to make the entrance level with the floor, size 20 × 30, with the hay-mow directly over it. Ten feet from the back end there is a partition, and two feet forward of that, another, enclosing a dead-air space, which will prevent all sudden changes of outside temperature from being felt within. Several days of warm or of very cold weather, will occur, before the difference is noticed by the bees. The walls are simply plastered, but should be laid in lime mortar to keep out rats and mice. The bottom is cemented thoroughly, and the top lathed and plastered. A tube four inches square is put in on one side near the bottom, for the admission of air, and another at the top, for its exit, both covered with wire-cloth to exclude vermin. A slide should be inserted in these tubes to regulate the supply of air.

Around the sides of such a room, arrange shelves at

proper distances, one above another, upon which to set the hives. If the cellar is very dry, some hives may be set on a board on the floor.

The bees should be allowed the benefit of flying out on all warm days at the end of the season, and should not be put into winter quarters until winter is positively at hand. Also, it is much better to remove them on a cool than on a warm day. It is an advantage, but not all-important, that each hive occupies its old stand when set out in the spring. To this end, they should be numbered, and when brought out, they can be placed where they are to remain.

HOUSING.

If the box hives are to be housed, place sticks an inch square (readily made by splitting a board of the right length), upon the shelves, and set the hives upon them, either inverted, or with ample ventilation at the top.

When the combs can be vertical, the best position is on the side. When all the permanent shelves are full, put in temporary ones until one hundred hives are deposited, which are as many as should be kept in one apartment. The movable comb hives should be set on sticks about half as large. But few of the holes need be open in the honey board, for ventilation, as the air can pass over the top of all the frames. When over one hundred are put in one room, they are apt to make the air too warm, in consequence of which the bees become uneasy, and some get out and are lost. If the least ray of light is admitted, they will go to it, and lose their hive. A small number—less than thirty hives—will not keep up the requisite warmth, and there is more need of confining the animal heat to each hive; this retains more of the moisture. It is better to reduce the size of the room in proportion to the number. A few in a large room will not do as well, as in one just large enough to hold them. They should be examined occasionally throughout the winter to see that all is right, but disturbed as little as possible. Perhaps when very cold, less air will be needed, than when moderate. Towards spring, they are less quiet, and you should take advantage of the first warm days after the middle of March to get a part of them out, so that the remainder will keep the temperature about the same as before.

SETTING OUT.

No matter if the snow is not all gone, if it has a crust and the day is warm, a bee will rise from it, just as well as from the bare earth. Eight or ten hives should be set out at once; after they have been out two or three hours, set out as many more. When all are taken out at one time, they are quite sure to mix, and unite with colonies where they do not belong. They are more particularly disposed to do so, when any stands have been changed or set in a strange place.

While the usual difficulty is to select a day sufficiently warm, it is possible, as the season advances, for it to be too warm. When the hive is first set out, the bees are intent only on getting out to fly. Those that have been out, are now on the lookout for plunder, and rush into the hives recently set out, and carry off the honey, before there is any guard established to defend it. When such days occur, the bees should not be taken from the cellar.

Should some hives, after all precautions, get more than their share of bees, while others are proportionally deficient, the best way is to simply change the hives, taking the strong one to the stand of the weak one, and the reverse. Take care that too many do not leave the strong hive and join the other, as will occasionally happen, in which case it will be necessary to return each to its own stand. Before making the exchange, ascertain if the weak one is not queenless.

If it be impossible to build such a cellar as the foregoing, one partially as good, is better than the open air for wood hives. It should be near the apiary, and if no side hill is within convenient distance, it may be made on the level, excavating for one half the depth of the room, if dry enough; or if not, make a room above ground. But it will require much more expense and trouble to make it answer as well as a cellar. The difficulty is in a lack of uniformity in the temperature, rather than in the inability to make it warm enough.

A few days of warm weather outside, will make bees uncomfortably warm within, and many will be wasted. To confine them to the hive does not avail much, as they will continue their endeavors to get out, until they worry themselves to death. Towards spring, this difficulty increases. A few bushels of snow or pounded ice laid on the floor, will do much towards keeping them quiet, till the time to set them out. A room, even above ground, should be made to obviate this difficulty in a great degree.

A BUILDING FOR THE PURPOSE.

Put up the frame of the desired size, board it inside and out, leaving a space of ten or twelve inches between, to be filled with some non-conductor of heat, such as sawdust or spent tan-bark. Then enclose this with another frame, leaving two feet on every side for a dead-air space in which, if necessary, snow or ice may be packed, which will have a tendency to keep the temperature more uniform. The inside can be arranged as already described.

ROOM IN THE DWELLING HOUSE.

Rather than the wooden hive should stand in the open air, I would appropriate some small room in the house;

make it dark, and secure as even a temperature as possible. Some bees will be wasted, yet not as many as on the stand.

BURYING BEES.

The conditions under which it is advisable to bury them, are sometimes found. A dry sandy soil is best. The pit should be dug where there is perfect drainage, and the hives should be surrounded on every side with straw, enough to absorb all the moisture from the bees, if not very much of that from the ground. The hive should be inverted, or laid on the side. As the bees produce some heat, it is not necessary to bury quite as deep as would keep potatoes from freezing. It is not advisable to put a great many in one pit, as there would be too much heat; also it would be necessary to take them all out at one time, very expeditiously, and the difficulty of their mixing, would be encountered. Make several pits, if a large number is to be wintered in this way. The labor is but little more, and there are certainly several advantages. A *very few* would winter better in a pit than in a room or cellar. In the first case they can be covered with earth, till properly warm, but in the other, they must keep themselves warm.

One object in protecting bees, is to save honey. The colder they are, the more they consume. The horse or ox consumes his food to replace the warmth that is thrown off on the cold air. The quantity seems to be regulated by the degree of cold; they refuse a portion of tempting provender in the warm days of spring, and greedily devour large quantities in the pelting storm. The farmer houses his cattle in winter on the score of economy. The same consideration should prevail with reference to his bees.

STRAW HIVES.

There are some bee-keepers who from some cause, cannot be induced to make a room or cellar available, and others who prefer to leave them on their summer stands. To such I would recommend the straw hive.

Not old fashioned conical shaped hives; although the objection to it, is simply the want of adaptation to improved bee-culture. We have all heard of the great success in wintering in "the old-fashioned straw hive," fifty or a hundred years ago. They were discarded, it is said on account of harboring the moth-worm, and inconvenience. Mr. Langstroth says: "Straw hives are warm in winter and cool in summer;" and again, "Hives made of wood are at the present time fast superseding all others." Notwithstanding this, I shall err greatly in my judgment if straw as a material for hives, does not in a great measure regain its former position in public favor. We now have straw hives of convenient shape, some of them covered by a patent, but that is chiefly on the manner of holding the straw. The proper degrees of heat and cold when most desirable, are great advantages, and can be obtained on principles long ago recognized.

PHILOSOPHY.

It is found that solid bodies are much better conductors of heat than porous ones. To illustrate: put on a rubber coat, or a woolen one, one impervious to air and water, the other freely admitting both; one conducts away the heat and retains the moisture, while the other retains the warmth, and allows the insensible perspiration to leave the body. A linen or cotton garment is a much better conductor of heat, than one made of wool. Perhaps this is owing to the fact that the fibres lie more compactly.

Air is a poor conductor of heat. We readily succeed in warming a room, but it is when the heated air can move from the fire, forming a current, and is replaced by cold air to become heated in turn. But confine the air in what we call a dead-air space, as is done in the walls of a house, or, if you please, within the interstices of a woolen fabric, and the heat passes off very slowly. I can readily conceive that straw, the leaves of the cat-tail flag, or broom-corn stalks, used as a material for bee-hives would act similarly as a non-conductor of heat, the thousands of tiny air-cells, being so many dead-air spaces to prevent the escape of the heat, and permit the passage of moisture. I speak comparatively, for some warmth will of course escape, but not so much by far as when a wood hive is properly ventilated.

It is thus apparent that a wood hive thoroughly water-soaked would conduct away the heat far more rapidly than when perfectly dry. In the one case the pores of the wood are filled with water, thus becoming a good conductor like a wet garment; in the other the pores are occupied with air, and the heat leaves slowly. The more numerous the air-cells, the slower it will pass. Hives made with double walls of boards, enclosing a dead-air space, do very well in regard to warmth, but do not dispose of the moisture with sufficient rapidity.

The moisture *must* be got rid of, and in no way can it be done so well, as by straining it through straw. Beside being advantageous for wintering, straw hives are superior in keeping the temperature warmer, and more uniform, throughout the spring, thus promoting early breeding and swarming. After the beginning of summer they do not seem to possess any special advantage over wood hives, further than that their combs are less liable to melt down. But the objection first raised by most persons—the harbor for the moth-worm—has not arisen in my experience. Out of a large number containing bees through the summer, not one has ever been injured in this way. However, I think it better economy to change the combs, bees, etc., to the wood hive for the summer, and back again in December. A straw hive will decay faster during use in summer, than in winter. Also it will receive a coating of propolis during July and August, that will render it less efficient in ridding itself of moisture. I say *less* efficient, because it is better than wood, at any rate.*

The top must certainly be removed to give place to the surplus boxes, and should be kept in the best possible condition for receiving the moisture, when used again the following winter. The top is much more important than the sides, and if the latter should become impervious to moisture and air, the bees would be warm enough to enable the top to absorb all the moisture before it would freeze. The operation of transferring from wood to straw hives may be performed at any time from October to January. The first cold snap in December is a suitable time.

Take the bees to a somewhat darkened room to prevent them from flying. Let the combs occupy the same relative position in the straw hive as in the other. Each comb, when lifted out, should have made in it a passage for the bees, three-fourths of an inch in diameter. It can be made in a moment with a knife, and is often important as a means by which the bees can pass from one comb to another, when in need of honey, and also to allow the queen to pass to different combs to deposit eggs, when the weather is not mild enough to permit her to go around the edges.

---

* Propolis can be removed from a hive if desired, by scalding, or crumbling, by striking or rubbing when very cold. Transferring the bees to some other hive meanwhile.

I like this method better than the coiled shaving recommended by Mr. Langstroth, because that cannot always be in the place most convenient for the bees. It is frequently in the midst of sealed honey.

The condition of the bees may be ascertained at any time by simply raising the top mat. As the hives sit near the earth, the snow may be allowed to drift around them. Should warm days occur before the snow is gone, when it is suitable for the bees to fly, it is not important to shovel it away from the entrance to allow them to issue. Remove the roof and slide the mat back a few inches; the bees will fly off and return without hindrance.

The walls of these hives are not as quickly warmed through by the rays of a winter's sun, as those of wood. Not until the whole surrounding air is mild enough for them to issue safely, do they become aroused, and desire to come out, and then it is usually safe to allow it.

STRAW TOP.

There are some apiarians who are unable to have the best and most profitable hive, and yet wish to winter their bees in the open air.

Perhaps a few such persons may be induced to make a hive that will admit of a straw top for winter. It would cost but little more to make a mat of straw, or flags, or even corn-cobs, to just suit the top of the hive.

Put bars or slats across the top to support the combs. The top board used in summer, may be held by screws, and is easily removed when desired. Such mat will absorb, nearly, if not quite all of the moisture, and the bees are in much better condition in consequence.

I will make some further suggestions to those who use the box hive, and have decided to take the chances of wintering on the summer stand with the least trouble.

SIMPLE BOX.

Entire success can attend those *only* who select *none* but the best stocks for winter, and secure for them the following conditions.

They *must* have air at all times and must be kept from freezing. The first condition will secure the last. If the bottom of the hive rests on the board, and there are but small openings at the bottom, and none at the top, all the moisture condenses on the combs and sides of the hives. A warm day melts it, and every thing in the hive is wet. Sudden severe weather freezes all solid. In this way, even *strong*, *heavy* stocks are lost. A special vent should be opened at the top to correspond with the bottom. A current of air passing through, will carry off the surplus moisture, and keep the combs comparatively dry, but a great deal of the heat that would be beneficial to the bees will go with it. This moisture is received in the cover of the honey boxes, which may, with benefit, be filled with hay, straw, or cobs to absorb it.

If the hive has no upward ventilation it should be raised at least an inch from the board, to give all possible circulation below; it will keep the lower ends of the combs dry, at least; but the upper ends may be a little frosty and moldy.

MICE.

To prevent the depredations of the mice when so raised, a strip of wire-cloth, a little more than an inch wide, surrounding the hive at the bottom, and held in place by a few tacks, will be effectual. The hole in the side should be covered with the same. It should be so put on in both places, that the bees may have room to pass at one edge. When thus guarded, the hives may remain under a snow drift for months, without danger from mice, smothering, or freezing. Indeed, bees could hardly spend the winter in a more desirable situation. In a few hours after the snow has covered them, it is melted for a space of four inches on all sides of the hive, and sufficient air circulates through it, for all their necessities.

SHADE.

It has been strongly urged to keep all hives out of the sun, without regard to the strength of the colony, because an occasional warm day allures the bees outside, when they get on the snow and perish. This is a loss, to be sure, but there is a possibility of inducing a still greater one by endeavoring to avoid the lesser. I have already said, that the second rate or poor stocks may occasionally starve with plenty of stores in the hive, on account of frosty combs.

If the hive is kept from the sun, in the cold, the periods of temperate weather may not occur as often as the bees exhaust the honey within their circle or cluster. But on the contrary, when the sun can strike the hive, it warms up the bees, and melts the frost more frequently. The bees may then go among their stores, and obtain a supply, generally, as often as needed. We seldom have a winter without enough sunny days for this purpose, but when such a one occurs, stocks of this class should be taken to a warm room, once in eight or ten days, for a few hours at a time, to give them an opportunity to get at the honey.

LOST ON THE SNOW.

As for bees being lost on the snow when flying out, I apprehend that not many more are lost there than on the frozen earth, that is, in the same kind of weather. I have seen them chilled and lost on the ground by hundreds, when a casual observer would not have noticed them; whereas, had they been on the snow, every bee would have been noticable at the distance of several rods. Snow is not to be dreaded so much as chilly air. Suppose that a hive stands in the sun throughout the winter, and the bees are permitted to leave when they choose, and a portion is lost on the snow; and that it were possible to number all that are lost by becoming chilled on the bare earth, throughout the season, the number lost on the snow would not be one in twenty. A person who has not observed closely during the damp and chilly weather in April, May, or even the summer months, has no adequate conception of the number. Yet I do not wish to be understood that those lost on the snow are of no consequence, by any means. On the contrary, a great many are lost that might be saved by proper care. But I would like to impress the fact that warm air is essential, and that crusted snow is as safe a footing for a bee, as frozen earth. Even melting snow is solid footing for a bee; it can and does rise from it, with the same ease as from the earth. Bees that perish on the snow under these circumstances, would be likely to be lost in any case.

The worst time for them to leave the hive is immediately after a new snow has fallen, because it will not sustain their weight, and they soon work themselves down out of the sun, and speedily perish. Should it clear off pleasant after a storm of this kind, a little attention will probably be remunerated. To prevent their leaving the hive at such times, a wide board should be set up before it, at least as high as the entrance in the side, to protect it from the sun. But if it grows so warm that the bees leave the hive when thus shaded, it is fair evidence, that it will do to let them sally out freely, except in case of a new snow, when they should be confined to the hive.

The hive may be let down on the floor-board, the passage in the side covered with wire-cloth, and made dark; raising, at night again, as before directed.

I have known hundreds of colonies wintered successfully without any such care, the bees being allowed to issue whenever they choose. Their subsequent health and prosperity proved that it was not altogether ruinous.

It has been recommended to enclose the whole hive by a large box set over it, and made perfectly dark; with means for ventilation, etc.

For large families it would do well enough, as would also some other methods. But I would rather take the chances of letting them all stand in the sun, and issue at pleasure, than to have the warmth of the sun entirely excluded from the medium sized families.

I never knew a whole colony to be lost from cold alone, but I have known a great many to starve, merely because the sun was not allowed to melt the frost on the combs, and give them a chance to get at their stores.

There are some extremely economical bee-keepers who urge the additional objection to allowing bees to stand in the sun, that "every time bees come out in winter, they discharge their excrement, and eat more honey in consequence of the vacant room." What an absurdity it would be to apply this principle to the horse, whose health, strength, and vital heat is sustained by the assimilation of food! The farmer is not to be found who would think of saving his provender by such means. That bees are supported in cold weather on the same principle, is indicated strongly, if not conclusively.

Is it not desirable, if what has been said on the subject of wintering bees is correct, to keep our bees warm and comfortable when practicable, as a means of saving honey?

To winter bees in the best manner, considerable care is required. Whenever you are disposed to neglect them, you should remember that one early swarm is worth two late ones, and that their condition in spring often determines which you shall have. Like a team of cattle or horses, when well wintered, they are ready for the season's work, but if neglected, they need a long time in which to recruit, before they are able to make themselves useful.

CHAPTER XXIV.

THE ITALIAN OR LIGURIAN BEE.

REPUTATION.

Much attention has recently been attracted to this new variety. The reputation given it by extensive and intelligent apiarians in Germany and other places, induced some of our citizens to import a few colonies some time in '59 or '60. As to who took the lead in this enterprise, there are various conflicting claims, which I shall not be able to reconcile.

IMPORTERS.

Mr. Mahan of Philadelphia, and Mr. Parsons of Flushing, N. Y., were among the first to disseminate them from the imported stock. Afterwards, Mr. Rose of New York, obtained these bees directly from their native Alps, and sent out large numbers. Those, to whom the first were sent, soon spoke in the highest terms of their superior qualities; representing them as fully sustaining the reputation they had gained before their arrival.

SUPERIORITY.

It was said that they were larger, more beautiful and hardy, more prolific, and more industrious than the common bees; that they swarmed earlier and more frequently, were less inclined to sting, more disposed to rob, and more courageous and active in self-defence! In corroboration of this, Dr. Kirtland of Cleveland, Ohio says,—I quote from Mr. Parsons' Circular,— "Their beauty of coloring, and graceful forms, render them an object of interest to every person of taste. My colonies are daily watched and admired by many visitors. So far as my experience has gone, I find every statement in regard to their superiority sustained. They will no doubt prove a valuable acquisition to localities of high altitude, and will be peculiarly adapted to the climate of Washington Territory, Oregon, and the mountainous regions of California." The Rev. L. L. Langstroth says, "If we may judge from the working of my colonies, the Italians will fully sustain their European reputation. They have gathered more than twice as much honey as the swarms of the common bee. This honey has been chiefly gathered within the last few weeks, during which time the swarms of common bees have increased in weight but very little. The season here has been eminently unfavorable for the new swarms—one of the worst I ever knew—and the prospect now is, that I shall have to feed all of them except the Italians." Mr. E. A. Brackett says, "My experience thus far satisfies me that the value of the Italian Bee has not been overrated. The queens are larger and more prolific; the workers when bred in comb of their own building are larger, and their honey-sacs larger. They are less sensitive to cold, and are more industrious. In all my handling of them, (and I have done so pretty freely, lifting the combs and examining them almost daily,) I have never known one to offer to sting."

WHEN FIRST OBTAINED.

The reports of many others were equally favorable; and no one has as yet reported unfavorably. I obtained queens of this variety from most of the parties who had imported them—from Mahan, Parsons, and Rose.

OBJECT.

I procured them, not because I expected to find them as perfect as represented, but because I wished to be able to express an opinion of them based upon actual experience. I wished also to verify by ocular demonstration many points in the natural history of the bee, with regard to which some of us, it is true, were already sufficiently well satisfied, but which at the same time were not so clear to the majority of bee-keepers; and this object the new variety would materially aid me to accomplish. But I had become so accustomed to look upon the wonders told of patent hives as extravagant and fabulous, and circulated only to victimize the ignorant, that when I heard the seemingly extravagant praises of the Italians, I very naturally put them in the category of humbugs pertaining to the bee. When obtained, if I had any bias, it was against them rather than otherwise. I was satisfied that among the number of bees in my apiaries, I would have greater opportunities for investigation and comparison than most of those who had fewer colonies. All of us who have had much experience know, that among colonies of the common bee, apparently equal in all respects, there will sometimes be a difference of onehalf in results. This shows that nothing reliable can be deduced from the experience of one or two seasons in small apiaries—say of a dozen stocks, half of them Italians, even if the latter surpass the common bee by fifty per cent., although one might very honestly conclude that the greater thrift of the Italians in such case was entirely owing to their superior qualities. We can hope to arrive at a just conclusion, only, by comparing the results of numerous trials in large apiaries.

Determined to get the best, if there was any difference, I obtained a queen from each party. The first one produced fully one-half of her workers of the native color. Having no faith in the purity of this one I did not dare to rear a queen from her. The next was not obtained till late in the season, owing to the management of certain interested *friends* at headquarters—of which friends most men are unfortunate enough to have more or less. At length, however, I obtained two, one of them in time to rear a few queens, the most of which were dark. The workers were fine. The next spring (1861) I raised a large number of beautifully colored queens. I had several stocks of hybrids from the queens reared the previous year, which had mated with the native drone. On these, with one drone queen, I depended for early drones for the young queens. I found that the drones from light-colored queens were much better marked than those from dark ones—indeed, a great many of the latter appeared no better than the natives. Yet, they had to be considered pure. Nearly all the queens raised at this time, before the appearance of the native drones, produced genuine Italian workers.

PECULIARITIES.

I now began to watch their peculiarities with considerable interest. I had two colonies nearly all changed, several hybrids, and a number in which I had just introduced the queen. I had about 60 native colonies, and all the Italians, marked with the yellow stripe, which would have made about three good swarms, in one apiary. White Clover was blossoming in abundance, and the Early Red, or June Clover, in small quantities. Here was a chance to see if they frequented the red clover more than the natives. I found nine Italians to two natives on this plant. The two exceptions might have been black hybrids.* I discovered some on a little ball of wax, made by the squeezing together by the hand, of bits of old, dry comb, that had been accidentally left in the sun. It was packed on their legs like pollen, and carried to the hive. I had never seen the native bees thus engaged. Here was another item to their credit, which, although of little account in itself, suggested that if they could turn to good account one stone rejected by the builders, they might also other and greater ones.

LONGER LIVED.

But more important than this, it soon became evident that they were longer-lived. Some time in October I deprived three ordinary-sized colonies of their queens and united them, giving them brood from an Italian queen. This brood occupied both sides of a comb some five or six inches square. It matured and a queen was produced. At this time there was about one Italian to fifty or one hundred natives. There being no increase except of drones, the queen proving barren of workers, the colony was pretty well reduced by the last of May. But the proportion of the Italians to the natives had been steadily on the increase. I now introduced a moderate family of natives, in order to continue the production of Italian drones. In a few weeks more, they again became reduced, worms appeared, and the colony was broken up. Not far from one-third of the remnant was Italians. Evidence of the strongest kind was here furnished, showing that they live longer than the natives.†

---

* This was important to me. If the honey from white clover would sustain 60 or 80 colonies, that from the red would sustain nearly as many more, and I could keep double the number in each year.

† It explains how a greater proportion of very weak colonies of Italians are increased into strong ones, than there are of the natives. Also how they retain their strength when all their combs are so nearly filled with honey, that but few cells are left for breeding.

ROBBING.

Their robbing propensities were also closely watched with the expectation of finding their appetite for marauding, insatiable. This propensity is indicated by their keen sagacity in scenting out any exposed honey that may chance to be in their vicinity. If standing uncovered on a table in the dining room, with the tempting avenue of an open window or door, the bees are quite sure to find it, especially at certain seasons, but the first one on hand is sure to be an Italian, notwithstanding nineteen twentieths of the apiary may be natives. Judging from this alone, we should conclude at once that they are unscrupulous robbers; and no doubt they are when there are colonies within reach, reduced to entire helplessness. But with me they were not half so troublesome as the black rascals. Whether I had no weak stocks to tempt them to begin, or whether they had a little principle of forbearance, I cannot tell. To see what they would do, I now had them standing promiscuously with the others throughout the yard. They were kept thus for two seasons with this object expressly in view. I thought it very likely that the wealth of stores which they were reputed to gain much more rapidly than the common bee, would be found to be composed in great part of plunder taken from their neighbors. But the idea had to be dismissed. In self-defence, they were vigilant and active. If a native approached the entrance of their domicile, he was seized and despatched without hesitation. Even an Italian venturing too near a strange colony was not favored. But I have kept my little boxes for rearing queens, successfully, though defended by only a handful of bees, till late in November.

The results of this season's experience were very satisfactory. I found the stocks of Italians, the hybrids as well as those in which the queens had been early introduced, averaging heavier than the others. Where the queens had been introduced, however, the season was well advanced before the bees were all changed, and such stocks therefore did not furnish a perfect test. Neither did the full blood colony from which I was taking brood to rear queens, since it was fed at different periods to promote breeding, and at the same time kept reduced to prevent swarming.

In the spring of 1862 I sold nearly all my pure queens, keeping only a few to breed from, and the hybrids.

DISPOSITION.

Long before this time, I had learned much about their *amiable* disposition. They had exceeded my expectations in so many particulars, that for a long time whenever they manifested any unusual ill-nature I found myself seeking some apology in the peculiar circumstances of the case. But I was at last reluctantly forced to admit that the Italian bees, especially the hybrids, were cross—not moderately so, but just as cross, it seemed, at times, as they knew well how to be. In the season of swarming for instance, to hive *some* swarms without protection would be perfect madness; others would be less irritable. In the season of honey, any time between ten o'clock in the morning and four in the afternoon, when the weather is fine, I have no difficulty in opening the hive to obtain brood, or for any other purpose. While at work they do not seem to notice much that is going on around them. Walking among them at such times seldom attracts attention. But when I would fasten up a colony that had been sold, and was now about to be sent away, I had to do it of course when the whole family was at home, usually in the morning; and at such times every bee would seem a warrior bent on driving me away. By the use of smoke I could drive them like the black bees among the combs and out of the way, but while the latter would be quickly and easily subdued, the former would return again and again, darting at my face like a shot, and not always without effect. They are remarkably quick. When I stand within two or three feet of a family of black bees, and see one start for my face, I can often avert it in time to prevent a sting; but he must be a skillful swordsman, who would thus parry the lightning-thrust of the Italian.

The results obtained during the summer of 1862 corresponded with those of the preceding year, and tended strongly to establish their superiority.

SWARMING.

Having a large proportion of hybrids, I had an opportunity to observe their swarming qualities. I found that they swarmed more, began earlier, and continued later, than the native bees. During the season of '63 there was a still greater difference manifest. The hybrids and a few pure ones, about seventy in number, constituted the whole apiary. Having no native bees in the yard with them—which of course would have furnished a more complete test—I had to compare them with others near by. Within three miles, in different directions, were six large apiaries, comprising nearly 300 colonies. If I except a few hybrids in one yard, the whole of them together failed to produce as many swarms as this one. Yet each of these six apiaries had the advantage of pasture, being located on the outside. The Italians began to swarm three weeks before the others. The first one, a hybrid, issued on the 20th of May, and a second from the same hive, on the 30th. As this was a season of but little honey, these two were put in hives containing combs and a little honey gathered the year before. By the 11th of July the old hive had again become full of bees, and a third swarm issued. The first swarm, hived May 20th, sent out a swarm June 30th, and a second July 9th; while the second swarm, hived May 30th, swarmed July 19th, making six swarms from one in a season. The other Italians did not swarm so excessively. The last three swarms got mixed with others and further trace of them was lost. The old hive and the first one from it, contained at the end of the season, strong colonies with ample stores for winter. The season for honey was one of the poorest that I ever knew. So little was obtained that but few of the natives could afford to swarm, and many that did so, failed to secure stores for winter, as did also some of the swarms. Some of the Italians did but little better in gathering honey. When it is not to be had, all must do without. But they gathered pollen, and reared brood, with thrice the energy of the natives. Swarms came out as late as 22nd of August, when scarcely any honey was to be collected. This propensity to swarm in such a season—I will not call it *over*-swarming, because in all cases, bees enough were left—was of no particular advantage. I mention it to show their perseverance in improving all possible conditions. It will be supposed that if they swarm thus in seasons of scarcity, a season of plenty would cause them to issue still more extravagantly. I have not found it so thus far; for this reason, as I suppose. The combs are quickly filled with honey, and brood is excluded. In spring, there are empty combs of course, and they fill up with brood, while the flowers yield little else than pollen. As soon as the first swarm leaves, which is usually in a season of honey, every bee that hatches leaves a cell that cannot be again occupied with an egg, within two or three weeks, which will allow the bees to fill it with honey; and by the time the young queen is ready to commence laying, her field of operations is limited to a very few combs near the bottom, not enough to admit of rearing bees for another swarm.

HIVE CROWDED WITH BEES IN COOL WEATHER.

On examining the hives at the commencement of cold weather, I found the whole colony packed into the small space at the bottom where the brood was hatched. A person not acquainted with the cause of their clustering so low down, would at once suppose that he had a prodigious family, when in reality the whole number would be no greater than in a colony of natives, where they were gathered half way to the top. It must be admitted that a colony with an excess of stores, is not in the best condition for winter, especially in the open air. Very likely the complaint will be made when this is the case, that the Italians do not winter well, even when "the hive was full of honey."

REMEDY.

The stores may be reduced, and the condition for winter improved, by dividing such colony at the proper season, and giving them empty combs for raising brood, or empty frames in which to construct combs. Both divisions will soon have plenty of breeding cells, and at the end of the season, will probably be stronger, than if confined to the few cells sometimes left for breeding in the full colony. If dividing would make them entirely *too weak*, it would benefit them greatly to remove several full combs, and replace with empty combs or frames. These bees are liable to excesses; when honey is scarce, they rear brood; when abundant, they gather too much for their own good. They will need supervision, and movable combs of some form are requisite. In giving my experience, I have given what may be considered as the general rule. There are exceptions in individual cases as with other bees. To single out and report either extreme is unfair; that which is true of the majority, is the only reliable criterion.

I have now related my observations on nearly all the points of character enumerated at the beginning, as well as some others. That they are less sensitive to cold, I am not yet prepared to say. That they are more prolific, is sufficiently proved. It is also clearly indicated that they are less so, in good honey seasons. It is also shown that they swarm earlier. Their disposition was found generally much worse than represented, yet under some circumstances very mild.

I think it very probable that many who obtain them will expect too much, and meet with disappointment. They may procure a colony that proves to be the exception to the rule, or the queen may be impure, producing nothing but hybrids. Although a half blood progeny will be much superior to the natives, the next remove will be so much reduced that they will not be a fair sample.

PURITY TO BE SECURED

I would advise all those who are disposed to try them, to purchase only of some reliable man, who will guarantee the purity of all he sells, so that if the first queen procured, should prove impure, it would be replaced by others, until a pure one was obtained.* The lowest price is not always cheapest. Bee-keepers have not as yet been able to decide on a reliable test of purity, a test that would detect the slightest mixture of native blood, with the genuine Italian. All admit that a yellow band must surround the abdomen of pure Italian workers; and that

---

* To change a colony of bees from the native to the Italian, it is only necessary to remove the native queen and introduce a pure Italian. She will at once commence laying eggs, and in about three months, the whole colony will be Italians.

the drones, a part of them, at least, must be marked irregularly, the band being somewhat scalloped.

VARIATION IN COLOR OF QUEENS.

Among the queens there is a great variation in color, some being even blacker than the natives, while the abdomen of others, is a beautiful yellow nearly its whole length. These are the marks, it is said, by which they are distinguished in their home in the Alps, where they are surrounded by a barrier of mountains impassable to the common bee.

It is said that they have existed there since the days of Virgil and Aristotle. If they existed at that time, as a superior variety, it would seem to be a mark against them that they have not become the predominant variety. That they should be indebted to the protection of surrounding mountains for their very existence is not much in their favor. It seems to be a law of nature that the poor and feeble shall be superseded by the better and stronger races.

As they are not a distinct species like the Stingless Bee of the tropics, but only a *variety* of the common bee,—as is proved by their mixing with them through all the grades,—I would suggest that they have grown into their present status through the influence of climate and surrounding circumstances, and that the impassable mountain enclosure has prevented all degrading alliances.

SUSCEPTIBLE OF IMPROVEMENT.

That bees may have changed from the common black to the brilliant specimens before us, in a few centuries or thousands of years, is indicated by similar changes in our domestic animals. We have the Pony, Cart, Farm, and Thoroughbred—horse: we have the Native, Ayrshire, Devon, and Durham, in neat cattle; the Newfoundland, Terrier, Hound, Pointer, and Poodle, among dogs, and all gradations of domestic fowls from the enormous Asiatic, to the tiny Bantam. All these differences have followed some adequate producing cause, and had we the whole genealogy in each case, we could doubtless go back to one original stock. Great changes are effected by selecting some point desirable to propagate, such as size, symmetry of proportions, or color, and breeding from such only as exhibit the desired qualities in the greatest perfection. The longer we breed in one direction, or the greater the number of generations that have exhibited particular qualities, the more we expect to find those points in the offspring, and the more the chances of their showing the original type, are diminished. But a few years since, a man conceived a fancy for breeding Shanghai fowls with short legs. He obtained by the first cross but a few specimens. Selecting the largest to breed from, he obtained a greater proportion of the mixed ones, and after a few generations, he had almost established a new variety, yet the bodies were not quite as large as he wished. By crossing again with the large ones, he obtained a few with short legs, three-quarter size, and by continued experiments, he was finally able to show these fowls with short legs, and bodies so little inferior in size as not to be distinguished by that alone.

Such examples of progressive improvement point out the way in which we can improve these bees in color, if in nothing else. We have only to breed from the best specimens, and as several generations can be matured in one summer, there should be rapid progress.[*] Until all the bees in a neighborhood are of this kind, there will be constant danger of mixing with natives. There must be continual vigilance, to discover and remove all such. It will not be long before all bee-keepers become satisfied of

---

[*] At present very many of the pure queens are dark colored, even when their workers are all handsomely marked. We should get rid of this feature.

the superiority of the Italians, before which time it cannot be expected that they will do much towards changing their stock.*

NEIGHBORS JOIN IN PURCHASING QUEENS.

Could all that keep bees in one neighborhood be induced to begin with the Italians at the same time, each would have but comparatively little trouble, but as this cannot be looked for, we must consider how a person must manage to preserve their purity when surrounded by the natives. There are two or three methods.

MIX THREE MILES DISTANT.

If the natives are three miles distant, there is but little risk. Should they be nearer, the young queens should be taken to some place at least that distance from them, so as to meet the pure drones before they commence laying.

COLONY TO FURNISH DRONES.

A colony to furnish the drones should be strong and must stand near the young queens. One for the purpose will rear many more than the usual number, by supplying it with extra drone-combs. Take out worker-combs and replace with drone-combs—taken one each from different hives,—leaving only worker-combs enough to maintain the strength of the colony.

As soon as the queen hatches and leaves the cell, the box containing her should be taken to this yard, and left there till she begins to lay, which will be in about eight days, more or less, when she may be introduced to the native stock. If there should be any chance for such queen to meet a native drone, which would be possible,

---

* During the past summer, (1864,) I succeeded in interesting my beekeeping neighbors, with one exception, within three miles, to Italianize their bees, and consequently expect to find but few hybrids among my young queens throughout the coming summer.

if one or two native stocks were within the prescribed limits, and it was desired to test her purity before introducing her, it would be necessary to wait till some of her workers hatch. If she has demeaned herself by a "mis-alliance" it will usually be manifested by some of her workers being black. If she is bright, and it is absolutely certain that she was purely impregnated, the bees left after she is removed, may be allowed to rear another queen from her brood; otherwise, give them brood from a queen that you know to be pure.

METHOD OF ITALIANIZING A WHOLE APIARY.

There is another method of changing an apiary where there is a large number, that may be less trouble than the foregoing. Those who are not fully acquainted with the natural history of bees, will be hardly able to comprehend it, yet I shall give it, and trust to their becoming familiar with the whole subject.

Queens enough can be reared in one summer to supply the whole apiary, no matter how many may be required, and if this is decided upon, take no pains to isolate, but rear all the queens at home, and let them meet the native drone. These will produce mixed workers, but pure drones.—The impregnation of a queen has no effect on her drone progeny.—Introduce a queen into each stock, and the next season, perhaps before, if done early, all the drones will be pure Italians. Now raise another set of queens from the original pure one; the drones of the yard being all pure, they can hardly fail to meet them, and consequently the second generation of workers will be pure. Occasionally, a queen will produce hybrid workers; as soon as discovered, remove her, and substitute another.

RAISING AND INTRODUCING QUEENS.

ARTIFICIAL QUEENS.

I term queens reared as about to be described, artificial queens, in contra-distinction to those raised by the bees in the process of natural swarming.

We will first see how it is to be done with the movable comb hive, for it is to be presumed that most who raise them will use this hive in some form, as without it, all the advantages will not be realized.

HOW TO REAR THEM.

Rearing queens artificially is done by enclosing a few bees, a pint or a quart, without a queen, with a small piece of comb containing larvæ or eggs. Make a little box, or miniature hive, large enough to hold three combs or more, four or five inches square. Suspend frames within, just as in the hive. Fit in them pieces of dry clean comb, and fasten with a bit of tin. Get a piece of comb containing eggs or larvæ, cut in this shape, two inches long by a little more than half an inch wide. Cut a piece the same size, except underside, out of the middle of one of the combs, and insert it thus, supported by each end. — See Fig. 34. — The bees will weld it fast in a few hours. The space cut out below, gives room to make the queen-cells, and they are quite sure to make them here. When the larvæ are just the right age, six or eight queens will sometimes mature in 10 or 11 days, at other times, in 16 or 18. But if the grub is over four days old, it is doubtful if it can be changed to a queen. This shape of the piece of brood comb is better than square, as it gives a

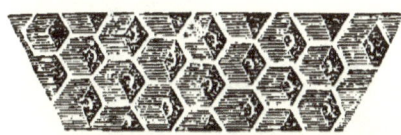

Fig. 34.—COMB CONTAINING BROOD FROM WHICH TO RAISE QUEENS.

chance to separate the cells should there be several. On a square piece they overlap each other, so that one cannot be cut out without spoiling most of the others. If you want to make the most of the cells you have, get ready another box with combs and bees, cut out a piece in the

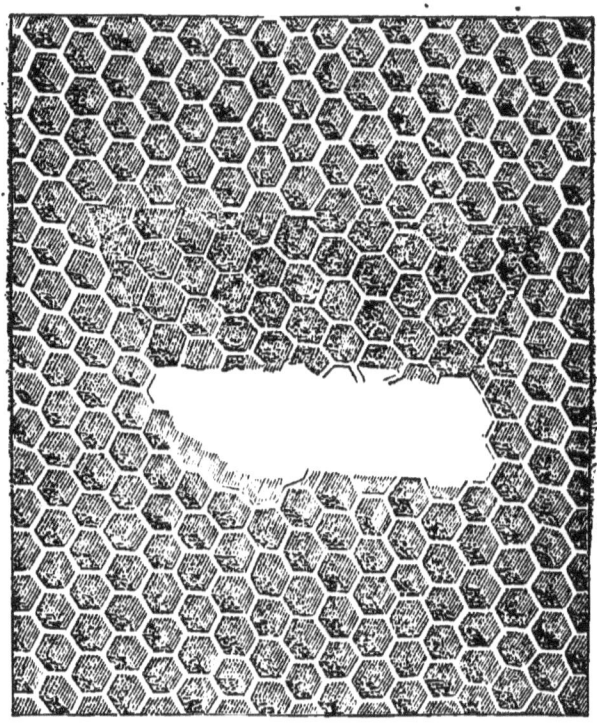

Fig. 35.—NO. 34 INSERTED IN A COMB READY FOR THE REARING BOX.

centre of the comb, and having carefully cut out a queen-cell, put it in this space, without bruising or rudely shaking it. The bees accept this instead of brood, and they will have a queen ten days sooner by the operation.

In about one-third of the cases, the bees will destroy such cell; the operation must then be repeated. It would insure safety in all cases, if the cell, with three or four bees from the rearing box, and a small quantity of honey in a cup of tin-foil or some convenient material, could be enclosed in a wire-cloth cage, and remain thus until the queen hatches, and they have become reconciled to her presence.

It will not do to postpone the removal of these cells for more than ten days if you wish to be sure of saving

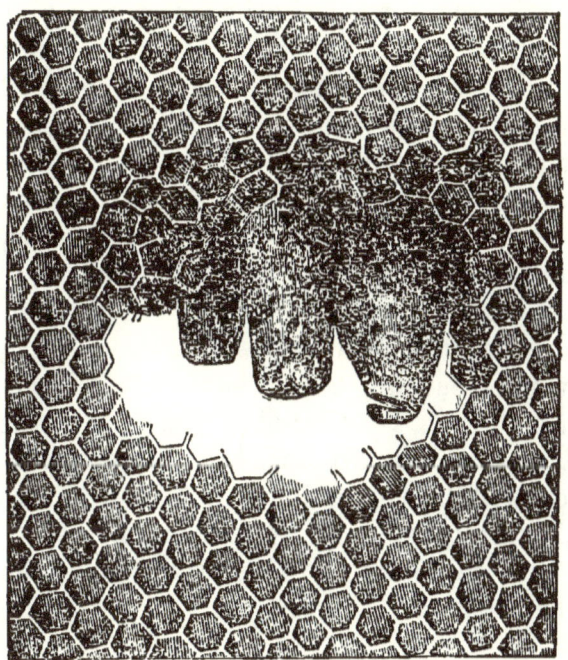

Fig. 36.—QUEEN CELLS MADE ON INSERTED COMB.

all the queens. I have known the oldest one to come out of her cell, and destroy all the others in ten days from the time they commenced rearing them.

HOW TO OBTAIN BEES FOR REARING QUEENS.

The bees to rear queens, should, when practicable, be obtained from hives at least a mile and a half from the place where the queens are to be raised. Take them from a strong colony. If from the box-hive, invert it and drive out a quart or two into an empty hive or box, look out the queen if among them, and put her back. If they are to be taken from the Movable Comb Hive, take out two or three combs and shake the bees off beside the box, into which they will run if it is set down with one edge raised a little, taking care all the time to not get the queen. Shut up the bees by tying a cloth over. Have an inch hole in the top of the box containing them, and when the small box, with comb, brood, etc., is all ready, set it over it, and the bees will enter at once. Not finding a queen, in a few hours they will commence rearing one or more, by converting common cells into queen-cells, and worker-larvæ into queens. If the bees have been taken from a colony at home, it will be necessary to confine them from thirty-six to forty-eight hours, otherwise they may return to the old colony. If taken from a colony at a distance, less time will answer. They should be fed when shut up, unless some of their combs are filled with honey. By obtaining the bees in the middle of the day while the old ones are out foraging, a large proportion will be young bees that have never left the hive, which are considered by many to be better for this purpose than old ones; at least they cannot know the way to their homes.

BLACK BEES AS NURSES.

Much has already been said relative to the propriety of having black bees as nurses to raise Italian queens. Some allege that they impart some taint to the young queen, which affects all her future progeny. Mr. Langstroth, who is entitled to as much confidence as any one, thinks it makes but little difference which variety rears the queen, provided there is a goodly number and plenty of honey. I have never been able to detect any difference. The largest and best queens are reared near the swarming season. I have observed that a larger proportion of dark and undersized ones are raised in cool weather in the fall.

I have raised nineteen in twenty beautiful queens from one mother, in July and August; and from the same in October, three out of four would be black and small.

Mr. Langstroth suggests that the cause of this lies in their being reared by weak colonies. After close observation, I have failed to discover much in confirmation of this view. Weak nuclei with abundance of honey, in warm weather, raise fair queens; strong ones in cool weather raise very poor ones.

BEST TIME TO OBTAIN BROOD.

From noon till 3 P. M. is decidedly the best time in the day to obtain the brood. While busy at work, the bees have not time to notice what is going on. Go to the hive, containing your best Italian brood, and take out different combs, till you find brood of the right age, and with a sharp knife, cut out suitable pieces. Have at hand some empty clean brood comb, from which cut pieces to exactly replace them. The bees will soon fasten them in place. In these cases it is well to have some smoke on hand, in case of emergency, but it is seldom necessary. Care must be observed not to allow brood or a queen-cell to become chilled. The rearing boxes, being small, will be affected by the changes of the weather, more than hives, and on some occasions will need some protection. Throw a blanket over the box, or take it in the house for the night.

It is unnecessary to raise queens before there are any drones to meet them. It is said by some that the drone should be at least two weeks old. About the fifth or sixth day after the queen leaves the cell, she issues for the purpose of meeting the drone; if successful, she commences laying about the eighth day. This rule, however, like others, is liable to exceptions. The queen may be safely introduced to the native stock, by taking the following precautions. A laying queen is received better than a virgin. The colony to receive her should be prepared a few days previously by removing its queen.

FINDING BLACK QUEEN.

She is most easily found in the middle of the day when the workers are abroad. If you can take out the frames without alarming the bees, your chances of success are more certain. Protect the face, and proceed without smoke. Pry loose the propolis fastenings, and raise off the top with the utmost caution, without jarring or thumping the hive sufficient to give alarm. Have an empty hive near, in which to put the frames as you take them out. Examine the centre combs, or those filled with brood first, when you can conveniently. Look over each one carefully; if not disturbed, the bees will be spread evenly over the surface, and her majesty is easily distinguished, when she can be readily taken up with the fingers. But if an alarm is raised, she is the most timid of any, and will get away in the corners of the hive, or among a mass of bees, where it will need close scrutiny to detect her.* In such case, you can either return the whole, and try it again some other time, or divide, putting one half the contents in the empty hive, and if possible the largest share of bees. Separate the combs in each, putting them only in the alternate spaces. If several hives are used, they can be still farther separated, which will be of some advantage. They become quiet in a few minutes, and the queen will venture out in sight. The bees on the comb with her will be quiet, while the others will manifest considerable uneasiness. This will direct your attention to the proper comb. To return the combs to the hive, in the relative position before occupied, number them before any are removed.

There is still another method of finding the queen which

---

* The Italians are seldom much alarmed, and the queen is easier found.

may be preferred. Make a box about one foot square, having one of its sides in the form of a door, with hinges, and a catch to hold it shut. Let another side be made of narrow slats or strips five thirty-seconds of an inch apart. These strips are better made of zinc, cut very accurately, but wood will answer. Shake the bees from the combs and let them enter this box, close the door, and sift the bees—shaking them lengthwise off the bars. The workers pass through the spaces, the queen, and drones, if any, are retained. If shaken in front of the hive, the bees enter as they drop out. The shaking confuses them so that they are not disposed to sting.*

The bees, when returned to the hive destitute of a queen, will at once commence operations to remedy the loss, by converting some of the worker larvæ into queens. In about seven days all the eggs left will have passed the period when it is possible to change them thus. Now take out the combs again, and cut off *all* royal cells that contain larvæ. The safety of the queen introduced, depends greatly on their entire removal.

The queen might be introduced in less time than this, and be accepted, at least without being stung immediately, yet being so different from the old mother, they are not always satisfied, and when they have means, will sometimes rear another, notwithstanding her presence. It is best to allow them no such opportunity.

INTRODUCTION OF QUEEN.

Introduce the queen thus. Secure her with two or three of her workers, and a little honey in a wire-cloth cage, and insert it among the combs. At the end of twenty-four hours, she may be let out among the bees, and when the foregoing preliminaries are observed—will be—

---

* When the hives from which the bees are taken for rearing boxes, contain black drones, they may be separated very readily by this method of sifting.

as far as my experience extends—well received. I have succeeded with equal satisfaction, and much less trouble with the following method, due to Mr. L. A. Aspinwall. It is simply to immerse the queen in a little honey— slightly warmed, if necessary—and drop her among the bees, which immediately commence licking her off, and forget that she is a usurper. This is so much less trouble that I prefer it, and consider myself very much indebted to him for the suggestion.

I would remark that if the Italian queen is introduced in the swarming season, when the bees are gathering abundance of honey, and the colony is strong enough to divide—making two—it would be well to do so, as it saves the trouble of looking up more than one queen for two hives, and you can find this one with much less trouble. (Directions for dividing may be found in the chapter on artificial swarms.)

ITALIANIZING THE BOX HIVE.

There are some who will still continue to use the box hive, who will wish to furnish them with Italian queens. It can be done, but involves a little more trouble, and the bees are kept a little longer without a laying queen; there is also a delay of some weeks, before you are certain of success. The process is as follows:

Drive out the bees, find and remove the queen, allowing the bees to return immediately. According to the rule, in twelve days they will mature a successor, and the first one that hatches will destroy the others before they mature. In eight days she will commence laying.

This must not be allowed, but as soon as it is certain that she has destroyed all her royal sisters, which will be in two or three days at farthest, and before she begins laying, she herself must be destroyed.—Any immature queens found about the entrance will indicate the destruction of her rivals. Drive out the bees again, find and kill the queen, and again return them. There is no possibility of their rearing another queen, and the Italian may be introduced in three days without much risk, if the usual precautions are taken.

Should the first queen that matures in such case, lead out a swarm instead of destroying her royal sisters, (which she would be likely to do, at any time near the swarming season,) it will be some days later before they are killed, and unless the colony is very strong, it would be best to return the swarm. Hive it, and set it near the parent stock till the next morning, then set the bees to running into the old hive, and secure the queen. When the piping entirely ceases, it may be taken as evidence that but one queen remains, and that it is time to operate.

If it is desired to introduce a queen into a stock that has swarmed it can be done on the same principle. The only *important* point is to secure the queen remaining after the destruction of the others, before she has begun to lay. Any hive that loses its queen by her coming out to meet the drone, may be supplied by simply taking the trouble to introduce one.

A neighbor has successfully introduced them to the box hive, at the beginning of the swarming season. As soon as a stock can spare a swarm, and before any queen-cells are finished preparatory to swarming, he drives out in the middle of the day a small swarm, and removes the old hive a few feet and places the new one on its stand. The old bees, that are acquainted with the old place, return there and make it strong. Two days afterward the young bees that are hatching readily accept of any queen that is given them. Old bees would be likely to destroy them when given under similar circumstances.

Queens may be introduced into the box hive by another process, in October or November, after the queen is done laying, or at any time when there is no brood in the combs from which to rear queens. Drive out the bees, remove the old queen, and return them. At the end of a week introduce the Italian, and all will be right. If there is any risk of eggs or larvæ remaining, keep the bees out of the hive for a week; keeping the box containing them with a little food, in the cellar or any safe place, till it is too late to rear a queen from their own means; then return them, and at the proper time introduce the queen.

I will give a method by which, with only one movable comb-hive, a small apiary may be Italianized in one season. Firstly, introduce an Italian queen into a colony occupying such hive. Drive out all the bees of some good stock into an empty hive, and set this on the stand. Take the hive from which the bees were driven, with its contents, to the stand of the one with movable combs. Lift out the combs and shake or brush the bees down by the box-hive, which they will immediately enter. Now take the movable comb-hive with contents to the other stand, and put that colony in it, and your colonies have simply traded hives, and each will carry on its usual operations, the same as if it had always been there. The one with movable combs can now be controlled. After a few hours, when the bees have become quiet, take out the combs, find and destroy the common queen.

In a week cut out all queen cells, and give them an Italian queen, and when she has filled the comb with eggs, four or five days after, this colony may be transferred also. Continue the process until all are changed. The cells cut out being Italian, may be put in the rearing boxes to hatch.

TRANSPORTING QUEEN.

A queen with a handful of bees can, with proper care, be sent safely one thousand miles by express. To pack her properly, have a box just a little larger in length and depth than one of the small frames in the rearing boxes; width about two inches inside. The bottom should be square; top the width of the box, and held by screws. The comb should be old and tough, and contain honey enough for the journey.

CHAPTER XXV.

PURCHASING STOCKS AND TRANSPORTING BEES.

QUALIFICATIONS FOR AN APIARIAN.

If the reader has no bees, and yet has had interest or patience to follow me thus far, it is presumptive evidence that he possesses the perseverance requisite to take charge of an apiary. He must, however, remember the inevitable anxieties and perplexities, and the amount of time that proper care requires, as well as the advantages and profit. But if he is disposed to try the experiment, some initiatory directions may very likely be acceptable.

LUCK.

The apparent uncertainty of success in bee-keeping has encouraged a general belief in the old tradition of "luck," and in no particular must the "fickle dame" be conciliated so much, as in the manner of obtaining the bees. Concerning this important operation, there seems to be a variety of opinions. One will assert that favor is secured by stealing one or two stocks to begin with, and returning them at some future time.

Another, a little more conscientious perhaps, says, that you must take them without *liberty*, but leave an equivalent in money on the stand.

A third assures us, that the only way to secure an effectual charm, is to exchange sheep for them; and a fourth affirms that *bees must always be a gift.* These methods have all been recommended to me, with gravity enough to make an impression.

But another method has been discovered, which works very well, and that is, when you want bees, go and buy them, and pay for them, in dollars and cents, or some other equivalent. And you need not depend on any *mysterious* charm, for success,—if you do, I can but predict failure. It is true that a few believers in "luck" will occasionally prosper, but it must be the result of accident, for where the true principles of management are not observed, how can it be otherwise? It is a saying with some that "one man can have luck but few years at once," and others, none at all, although he tries the whole routine of charms. Thirty years ago, when my respected neighbor predicted a "turn in my luck, because it was always so," I could not understand the force of the reasoning, unless it belonged to the nature of bees to deteriorate, and consequently run out. I at once determined to ascertain the truth for myself.

I could understand how a farmer would often fail in raising his crops, if he depended on chance or luck, instead of upon the fixed principles of nature. It seemed to me quite possible that the same reasoning would apply to the culture of bees. I observed that in good seasons the majority of bee-keepers were "lucky," and in poor seasons, the reverse; and when two or three of the latter occurred in succession, they always "lost their luck." It was evident then, that if my bees could by any means survive the poor seasons, they would do well enough in good ones. The result has given me but little reason to complain.

My advice therefore, is, that reliance be placed on

proper management, alone, and that all superstitious notions of propitiating some mystic power be thrown to the winds.*

It is quite common for beginners to take bees "on shares" as it is termed; it is a cheap way to begin, and there is no risk of loss in capital.

The general rule is this: one or more stocks are taken for a term of years, the person taking them, finding hives and boxes, and bestowing the necessary care, and returning the old stocks to the owner, with half the increase and half of the annual profits.

Yet, if bees prosper, the interest on the money paid for them is a mere trifle compared to the value of the increase, and there is the same trouble. On the other hand, the owner can afford to take care of a few hives more, for the half of the profits which he has to give, if another takes them.

There are yet a few persons who refuse to *sell* a colony of bees, because it is "bad luck." There is often a foundation for this notion.

Suppose a person has half a dozen hives, three extra good—the others, the opposite. He sells the three good ones, for the sake of the better price; there is but little doubt but his "luck" will go too. But had he sold the poorer ones, the result would doubtless have been very different.

But sometimes apiarians have more bees than they wish too keep, and such are the ones of which to buy. Purchasers seldom want any but first-rate stocks—such are generally cheapest in the end.

---

* I receive scores of letters, detailing the continued success of the writers, till they can count their colonies by hundreds, arising from the adoption of a common sense method of management.

PURCHASE THE BEST.

Firstly then, select first-class stocks; it will make but little difference whether they are obtained in fall or spring, if winter management is understood. I have already said that the requisites for winter were, a numerous family and plenty of honey, and that the cluster of bees should extend through nearly all the combs.

AVOID DISEASED STOCK.

To avoid diseased brood, make your purchases, if possible, in an apiary where it has not made its appearance. There are some who have lost bees from this cause—and yet were totally ignorant of the fact. It is well therefore, to inquire if any stocks have been lost, and trace out the cause, being careful not to mistake the immediate occasion of the loss, for the primary one—which may be disease.

OLD STOCKS NOT OBJECTIONABLE.

If you are satisfied that there is no foul brood, you need not object to stocks two or three years old, they are as good and sometimes better than others, especially if they have swarmed the season previous, because such always have young queens, which are said to be more prolific than the old ones, which are nearly always found in first swarms.

When no apiary from which to purchase can be found except those in which the disease prevails, you cannot be too cautious in making a selection. It would be safest in this case to take none but young swarms, as it is very uncommon for any to be affected the first season.

Old stocks are as prosperous as any, as long as they are healthy, but they are more liable to become diseased than young swarms.

If you are not allowed to take all young stocks, examine them in pretty cold weather, as the bees will be farther up among the combs, and give an opportunity for inspecting them. About November, all the healthy brood will be hatched. Sometimes a few young bees may be left that have matured, and have been chilled by sudden cold weather, but these are not diseased— the bees will remove them the next season, and no bad results will follow. In warm weather, a satisfactory inspection can be made, only with the use of smoke. Be particular to reject all that are affected with the disease in the least; do without, rather than begin with such. (A full description of the disease has been given in Chap. xiv.)

A neighbor once purchased thirteen hives; six were old ones, the others swarms of the previous season. He probably knew nothing of foul brood, and as the old hives were heavy, he thought them good, but five of the six were badly affected. Four were a total loss, except the honey; the fifth lasted through the winter, and then had to be transferred. He had flattered himself that they were obtained very cheaply, but when he estimated the cost of the good ones, he found no great reason for congratulation.

Another point is worthy of consideration: endeavor to get hives as near the right size as possible, (viz.) 2000 cubic inches,—better too large than too small. If too large, they may be cut off, leaving them the proper size, although this often makes the shape ungainly. But as the shape probably makes no difference in the prosperity of the bees, when extremes are avoided, the appearance is the principal objection.*

---

* A hive may be cut off with very little trouble in a cold day. Turn it over— the bees will soon find it too cold to venture out—mark it the right size, and saw it off. Lift off the piece, and trim off the combs even with the bottom of the hive. Use smoke to drive the bees from the ends of the combs.

TRANSPORTING BEES.

In transporting your bees, avoid, if possible, the two extremes of very cold, or very warm weather. In the latter the combs are so nearly melted, that the weight of the honey well bend them, bursting the cells, spilling the honey, and besmearing the bees. In very cold weather, the combs are brittle, and easily detached from the sides of the hive. When it is necessary to move them in winter, they should be put up an hour or two before starting. The agitation of the bees on being disturbed will create considerable heat, which imparted to the combs, will make them less brittle.

Have ready some carpet tacks, and pieces of thin muslin about half a yard square. Invert the hive, put the cloth over, neatly folded and fastened with a tack at the corners, and another in the middle of each side. Crowd the tack in about two-thirds of its length; it is then convenient to pull out when required.

If the bees are to be taken some distance, and must be confined for several days, the muslin will hardly be sufficient, and wire-cloth must be substituted. New comb will break more easily than old. Probably the best mode of conveyance is in a wagon with elliptic springs. A wagon without springs is bad, especially for young stocks; yet I have known them to be moved safely in this way with care in packing hay or straw under and around them, and careful driving. When there is good sleighing, a sleigh will answer very well, and some prefer this method of transportation.

Whatever conveyance is employed, the hive should be inverted. The combs will then rest on the top, and are less liable to break than when right end up, because in the latter case the whole weight of the combs must come upon the fastenings at the top and sides, and these are easily broken.

It is considerable trouble to prepare the movable comb hive to be turned over, yet for long journeys, it is absolutely necessary. Put sticks on each side of each comb, in about two places, to hold it steady, (see directions for transferring combs in Chap. xix,) then lay on the top of the frames, cross-wise, thin strips to hold them in place—and fasten on the honey board with screws.

Turn the hive over, and cover the bottom with wire-cloth. With proper care they may be sent by Rail Road one thousand miles. I can devise no convenient way of fastening the combs in hives that have permanent bottom-boards, such as Mr. Langstroth and some others use. Some other patent hives, like Mr. Hazen's, cannot be inverted for transportation. Such must of necessity be carried right side up.

I sometimes transport movable comb hives in this manner for very short distances, but with much fear of breakage. When I send off a colony of Italians, I dare not risk them thus.

When bees are moved, thus inverted, they will creep upward; in stocks part full, they will often nearly all leave the combs and get upon the covering.

In a short time after being set up, they will return, except in very cold weather, when a few will sometimes freeze, consequently, they should be put in a warm room for a short time.

After carrying them a few miles the disposition to sting is generally gone. When bees are confined in moderate weather, they manifest a persevering determination to find their way out, particularly after being moved, and somewhat disturbed. I have known them to bite holes through muslin in three days. The same difficulty is often experienced in attempting to confine them to the hive, by cloth, when in the house in winter.

Should any combs become broken, or detached from

their fastenings, by moving, rendering them liable to fall when set up, the hive may remain inverted on the stand, till warm weather if necessary, and the bees have fastened them, which they will do soon after commencing work in the spring. If they are so badly broken that they bend over, rolls of paper should be put between them to preserve the proper distance, till secured. While the hive is inverted it is essential that there is a hole in the side, through which the bees may work. A board should fit closely over the bottom, and be covered with a roof to effectually exclude all water, etc. When they commence making new combs it is time to turn the hive right end up.

CHAPTER XXVI.

CONCLUSION.

In conclusion I would say that the apiarian who has followed me attentively, and has added nothing of value to his stock of information, possesses an enviable experience.

It has been said, that "three out of five who commence bee-keeping must fail;" but we must suppose that the fault arises from ignorance or inattention, and is not inherent in the bees. To the beginner, then, I would say: if you expect to succeed in obtaining one of the most delicious of sweets for your own consumption, or its equivalent value in dollars and cents, you will find something more to be requisite than merely "holding the dish to catch the porridge." "SEE YOUR BEES OFTEN," and and know at all times, their actual condition. This one precept is worth more than all others that can be given;

it stands at the head of all the duties of the apiarian. Even the grand secret of successful resistance to the worms "KEEP YOUR BEES STRONG" is subordinate to this. With proper and persevering application of the above motto, you cannot fail to realize all reasonable expectations. Avoid over-anxiety for a rapid increase; be satisfied with one good swarm from a stock annually—your chances of future success are better than with a sudden increase of numbers. You will probably be obliged to discard some extravagant ideas of profits from the apiary. Yet you will find one stock trebling, perhaps quadrupling its price or value, in products, while one beside it does nothing. In particularly favorable seasons your stocks collectively will yield a profit of one or two hundred per cent,—in others hardly make a return for trouble. I have known the proceeds of a single colony in one season to amount to $35.00; and an apiary of ninety stocks to produce over $900, some of which added not a farthing to the amount. A bee-keeper in an adjoining county reports a profit of $1,800 from one hundred and thirty hives in a single season. The proper estimate can be made only after a number of years, when, if they have been judiciously managed, and your anticipations have not been too extravagant, you will be fully satisfied.

I do not wish to induce any one to begin bee-keeping, and relinquish it in disgust and disappointment. But I would encourage all suitable persons to try their skill in bee-management. I say suitable persons, because there are many, very many, not qualified for the charge.

The careless, inattentive man who leaves his bees unnoticed from October till May, is the one who will be likely to complain of want of success.

Whoever cannot find time to give his bees the needed care, but can spend an hour a day in gossiping at the neighboring bar-room, is unfit for this business. But how can he, who has a home, and finds his interest divided between that and the idle attractions of the tavern, and wishes to withdraw from unprofitable associates, employ his time with a better prospect of success than in the care of bees? They make ample returns for every attention.

And the gain may not be altogether pecuniary. A great many points in their natural history are yet undiscovered, and the truth of many others disputed. Would it not be a source of satisfaction to be able to contribute a few more facts upon this interesting subject, and thus hold a share in the general fund of scientific knowledge?

Granting all the mysteries pertaining to their economy to be discovered and elucidated, precluding all necessity of further investigation, would the study be dry and monotonous? On the contrary, the daily verification of established facts would be so fascinating and instructive, that we could not avoid a sentiment of pity for the condition of that man who finds gratification only in the gross and sensual.

It has been remarked that " he who cannot find in this and other branches of natural history a salutary exercise for his mental faculties, inducing a habit of observation and reflection, a pleasure so easily obtained, unalloyed by any debasing mixture—tending to expand and harmonize his mind, and elevate it to conceptions of the majestic, sublime, serene and beautiful arrangements instituted by the God of Nature, must possess an organization sadly deficient, or be surrounded by circumstances indeed lamentable." I would recommend the study of the honey-bee, as one best calculated to awaken the interest of the indifferent. What can arrest the attention like their organism—their diligence in collecting stores for the future —their secretion of wax and formation of it into structures with a mathematical precision astonishing the profoundest philosophers—their maternal and fraternal affection in regarding the mother's every want, and assiduous care in nursing her offspring to maturity, and their unaccountable display of instinct in emergencies, filling the beholders with wonder and amazement? The mind thus contemplating such wondrous operations, cannot avoid looking beyond these results to their Divine Author. Therefore let every mind that receives one ray of light from nature's mysterious transactions, and is capable of deriving the least enjoyment therefrom, pursue the path still inviting onward.

Every new acquisition will yield an additional satisfaction and renewed courage for the next attempt which will be made with a constantly increasing zest; and he will arise from the contemplation, a wiser, better, nobler being; far superior to those who have never looked beyond mere animal gratifications.

Is there in the whole circle of nature's exhaustless storehouse, any one science more inviting, more refining, and more exalting than this?

INDEX.

A

Acer rubrum............................ 78
" saccharinum...................... 79
After-swarms, How issuing............178
" Queens of................179
" Leaving hive..........179
" Not choosing weather..179
" Size of, etc..............175
" To return them.........180
Alder, Common or Candle.............78
Alnus serrulata........................ 78
Alsike or Swedish White Clover....... 81
Althæa rosea........................... 88
Ants...................................232
Aphis.................................. 85
Apiarian, Qualifications for..........333
Apiary, Italianizing the..............322
" Location of....................100
" Location marked...............102
" Should not be moved..........102
Asclepias Cornuti..................... 82
Aspen................................. 78

B

Basswood..........................83, 99
Bee, Italian or Ligurian...225, 308, 311, 318
" " Disposition of............314
" " Purity of..................318
" " Robbing Propensities of..313
" " Swarming of...............315
Bee-bread in drone cells.............. 94
Bee-charms...........................225
Bee-houses...........................111
" Unprofitable................107
Bee-moth.........................129, 234
Bee-pasturage......................... 76
Bees, Age of.........................201
" Anger of........................221
" Battles of......................121
" Before young commence labor.. 26
" Black, for nurses..............326
" Burying........................299

Bees, Driving in cold weather.........207
" Enemies of.....................228
" Equalization of................120
" Examination of................149
" Feeding of....................122
" For raising queens.............325
" Housing.......................296
" How they attack...............223
" " to get rid of...............141
" Improvement of................319
" Injuring grain................. 89
" In moth-webs..............237, 238
" Issuing.......................153
" Killing.......................268
" Lost on snow..................305
" Manner of feeding...124, 126, 127, 129
" Nature of..................... 20
" Necessary to insure a crop..... 91
" Necessities of................. 48
" Paralyzing....................271
" Protection against............227
" Rough treatment of young...... 25
" Sagacity of...............263, 264
" Starvation of, in winter......287
" Sting of......................226
" Swarming of..................146
" Terms applied to young........ 27
" Transferring............259, 260, 262
" " advantages of..............294
" Transporting..............333, 338
" Warm room for................291
" Water for.................252, 292
" When boxes are taken off......141
" When no swarm issues.......... 35
" Wintering.....................284
" " after scarcity of honey 274
" " Cellar for................ 295
" " Straw-hive for............ 73
" With dysentery................287
Bottom board, Inclined................ 50
Box, Making holes in when full.......137
" Simple........................304
Boxes, Advantage of glass............139

INDEX.

Boxes, Putting on and taking off.......135
" Too easy of access...........138
" To prevent queen entering....139
" Transferred................193
" When to take off............140
Breeding and physiology........... 22
" In large and small colonies.. 23
Brood, Best time for..................327
" Diseased...................210
" " cause and remedy.....212
" " description of.....211, 216
" " examination of........218
" " Mr. Wagner's view of..217
" When they begin to rear....... 22
Buckwheat......................89, 99
Button-ball........................ 88

C

Cat-bird..............................230
Catnip.............................. 80
Cells, Containing honey for daily use.. 95
" For drones.....................251
" For rearing queens............. 29
" Uniformity of.................252
Cephalanthus occidentalis............. 88
Chamber hive........................ 49
Cherry, Wild........................ 79
Chickens eat drones..................230
Chrysalis........................... 27
Clover.......................80, 81, 99
Clustering bushes...................101
Colonies, Deserting when destitute....124
" Inclined to rob..............267
" Selecting for winter..........266
" Size of....................285
" When to feed................124
Conclusion..........................340
Comb, Commencement of...............217
" Constructed as needed........ 96
" Crooked.....................250
" Melting of..................254
" Removing...................279
" Straight....................250

D

Dandelion.......................... 79
Diervilla trifida................... 81
Disease............................210
" Cause and remedy............212
" Description of..............211
" Manner of spreading..........216
" Mr. Wagner's view of........217
Diseased brood, Knowledge of........220
Drone and worker-combs..............191
Drone cells........................251
Drone-comb.........................195
Drone-combs, Too many..............192

Drone eggs, Theories about unimpregnated..................89, 40
" " When laid.............. 36
Drone-layers......................36, 37
Drones............................. 19
" Age of..................... 20
" Colony for..................321
" Destroyed before swarming.... 33
" Eaten by chickens............230
" Needed.....................197
" Number of.................. 35
" Theories relative to.......... 36
" When met by queen........... 29
" When reared................. 33
" Why sometimes killed.........152
Dysentery among bees................237

E

Eggs, Number laid by queen........30, 31
" Of drones, When laid........ 36
" When they hatch............. 25

F

Feeding, A last resort..............122
" Care in...................123
" Manner of.................126
" Object of.................127
" Promiscuous...............129
" When best................277
" Refuse honey,.............281
Flowers of fruits................... 79
" Yielding pollen first.......... 78
Fruit flowers, Important............. 79
Fumigator..........................271

G

Glass boxes, Advantage of...........139
Grub.............................. 27
Guide-comb......................... 64

H

Hamamœlis Virginiana................ 86
Hive, Accessibility of..............265
" Chamber..................... 49
" Changeable.................. 51
" Common box...........58, 61, 304
" Dividing.................... 50
" Farmer's.................... 57
" In cold weather.............317
" Italianizing the box........330
" Moth-proof.............54, 134
" Movable comb................ 66
" " How made........ 68
" " " To use...171, 191, 194
" Non-swarmer................. 55
" Observatory................. 75
" Of straw, for winter......... 73

1*

346 INDEX.

Hive, Proper size of............ 59
" Setting out..................297
" Suspended.................. 50
" Top of, not fastened............ 62
" Ventilating................... 54
Hives................................ 46
" Best cover for..................107
" Certificates and premiums for.... 47
" Cheap stand for..............104
" Desirabilities of................ 65
" Discovery about............... 49
" Furnished for trial............ 48
" Housed for winter..............167
" Making holes in when full......293
" No patent for.................. 46
" Of different colors.............109
" Principles of................... 49
" Remedy when crowded.........317
" Setting out...................297
" Shade for....................305
" Should be ready...............154
" Space between................103
" Straw........................300
" Too high.....................105
" With inclined bottoms.......... 50
" Worms in.....................243
Hoarhound........................ 80
Hollyhock......................... 88
Honey-boxes...................... 63
"   " Transferred..............193
Honey and wax, Straining............279
" Best season for.............. 96
" Discharging................. 94
" Distance bees will go for......100
" Feeding refuse...............281
" From buckwheat.............. 89
" From one swarm.............250
" In boxes in the hive.........192
" In cells for daily use.......... 95
" Principal sources of.......... 77
" Substitute for................ 77
" The first..................... 79
" To secure from worms........143
Honey-dew........................ 85
Honeysuckle, Bush................. 81

I

Introductory........................ 17

K

King-bird, A word for...............229

L

Larva............................. 27
Laying............................ 24
Leonurus Cardiaca.................. 80
Leucanthemum vulgare.............. 81
Linaria vulgaris................... 81

Linden............................ 84
Liriodendron Tulipifera............ 83
Locust............................ 79

M

Making box-hives.................. 61
"   honey-boxes................ 63
"   movable comb-hives......... 68
Mallows........................... 88
Malva rotundifolia................. 88
Maple, Red........................ 78
"   Sugar........................ 79
Marrubium vulgare................. 80
Metheglin and vinegar.............281
Mice..............................304
Mice and rats.....................229
Mignonette........................ 88
Milkweed, Singular fatality of..... 82
Moth..............................234
" Destruction of.................241
" In hives......................243
" Remedies for..................244
" Where deposits its eggs.......236
Moth-proof hives..............54, 134
Moth-worm, Destruction of........129
"   Found in best stocks......129
"   Fears the bees............130
"   How destroyed.............131
"   In centre of comb.........237
"   In old stocks..............239
"   Size of....................240
"   Troubles small colonies...181
Motherwort........................ 80
Mustard........................... 84

N

Nepeta Cataria.................... 80
Nymph............................. 27

O

Ox-eye, Daisy..................... 81

P

Pasturage for bees................ 76
Physiology and breeding........... 22
Plant-louse....................... 85
Pollen, flowers yielding first..... 78
" Manner of discharging......... 94
" Manner of packing............. 77
" Substitute for................ 76
" Two kinds in one cell......... 92
Populus tremuloides............... 78
Propolis..........................256
"   Abundance of...............258
"   Wax instead of.............257
Pruning......................205, 209
Prunus serotina................... 79
Pupa.............................. 27

INDEX.

Q

Queen, Age and office of............... 18
" Description of................... 17
" Finding black.................328
" Introduction of................329
" Leaves hive to meet drone..29, 196
" Maturing of..................... 27
" Number of eggs will lay....... 30
" One destroys others............184
" Piping of........................176
" Presence of...............32, 93, 331
" Question about her leaving....196
" Regard of workers for.......... 18
" Similarity of eggs with worker.. 28
" Time to lay eggs................193
" Transporting the...............332
", When the old leaves........35, 153
" Young takes place of the old
  one............................. 35
Queen-cells, Introduction of..........190
" Artificial.....................323
" Black bees nurses for..,.......326
Queens, cells for rearing............... 29
" Destroyed before swarming... 33
" Laying........................ 24
" Loss of................195, 200, 203
" Loss of and remedy........201, 203
" Mixing of.....................321
" Obtaining bees for............325
" Ovaries of.................... 41
" Raising and introducing......323
" Reared in swarming hives.... 33
" Replacing....................109
" Time of leaving...............199
" To obtain bees for............325
" Variations in color of.........319
" When lost....................198

R

Rats and mice......... 229
Red Clover............................... 80
Red Raspberry........................... 80
Removing combs........................279
Renewing combs........................209
Reseda odorata.......................... 88
Rhus glabra............................. 84
Robbers, when to look out for.........116
" First indication of.............117
" Remedies for..................118
Robbing, Not understood..............113
" Difficulty in deciding.........117
" Weak colonies in danger of..115
Robinia Pseudacia....................... 79
Rubus strigosus.......................... 80
Rumex Acetosella....................... 80

S

Salix Vitellina.....................79, 85
Shade................................305
Silkweed, Singular fatality of.......... 82
Sinapis nigra......................... 84
Skunk-Cabbage 73
Smoker..............................224
Snap-dragon.......................... 81
Sorrel................................ 80
Spiders..............................234
Stand for hives, Cheap................104
Sting of bee.........................226
Stings, Remedy for..................227
Stocks, Causes of weakness...........208
" How many to keep............ 97
" Old not objectionable..........336
" Purchasing....................332
" Requisites of..................267
" Uniting.......................275
" Uniting poor..................269
Straining honey and wax..............279
Sumach.............................. 84
Swarm-clusters......................155
Swarm, honey from one..............250
Swarming...........................146
" How to do it..................155
" Indications of............148, 154
" of Italian or Ligurian bee...315
" Preparations for........20, 150
" When commences............147
Swarms, After, uniting................182
"   " troublesome...........182
"   " rule for................183
"   " how issuing...........178
"   " number of queens of..179
"   " not choosing weather..179
"   " leaving parent hive...179
"   " to return them........180
"   " size of, etc.............175
" Artificial....................185
"   " perplexities with....185
"   " work well...........186
"   " first experience with.186
"   " how to make........ 188
"   " placing stands for...189
"   " queen-cell for........190
"   " drone-combs for...191
"   " with movable combs
                        ....191, 194
" Choosing weather.........174, 175
" Clustering bushes for..........161
" Driving in cold weather......207
" First enough..................165
" How far they will go.........164
" How to divide................167
" How to keep separate........166
" Loss of, by flight............ 102

Swarms, Second, size of 175
"   "   when expected.......175
"   Selecting a home.............164
"   Shade important for..........160
"   Should all be made to enter..159
"   Sometimes return............173
"   Times of issuing............. 177
"   To divide in movable comb-
           hives........................171
"   To put in movable comb-
           hives........................160
"   United....................270
"   Water necessary for.........252
"   When on the wing............167
"   When they issue..............151
Swedish white Clover or Alsike....... 81
Symphoricarpus fœtidus............... 78

T

Theory, Mr. Wagner's, of drone-eggs... 39
"   Mr. Harbison's, do............. 40
Tilia Americana....................... 83
Toad..................................231
Toad-flax............................. 81
Transferring bees....................259-265
Transporting bees.....................333
"   queen......................332

V

Vinegar and metheglin................281

W

Warmth..........................284, 285
Wasp, black............................232
Wax....................................245
"   Abundance of.....................258
"   Honey consumed for.............. 53
"   How obtained....................246
"   Instead of propolis..............257
"   Making 282
"   Wasted......................251, 282
White clover.......................... 80
White-wood........................... 83
Witch-hazel, Unusual secretion of..... 86
Willows..............................78, 85
"   Golden 79
Wintering bees.....................284, 295
"   "   Building for...........298
"   "   Straw-hives for......... 73
Worker and queen-eggs................ 28
Workers.............................. 19
"   Age of......................... 19
Worm, Destruction of..................241
"   In old stocks..................239
"   Size of........................240
Worms, How they get in...............144
"   In centre of comb.............237
"   In hives......................243
"   Remedy for.................145, 244
"   To secure honey from.........143
Wren, Box for........................185

THE
MODERN HORSE DOCTOR:

CONTAINING

PRACTICAL OBSERVATIONS

ON THE

CAUSES, NATURE, AND TREATMENT

OF

DISEASE AND LAMENESS IN HORSES

EMBRACING

THE MOST RECENT AND APPROVED METHODS, ACCORDING TO AN
ENLIGHTENED SYSTEM OF VETERINARY THERAPEUTICS, FOR
THE PRESERVATION AND RESTORATION OF HEALTH.

WITH ILLUSTRATIONS.

By GEORGE H. DADD, M. D.,
VETERINARY SURGEON,
AUTHOR OF ANATOMY AND PHYSIOLOGY OF THE HORSE, AND THE REFORMED CATTLE DOCTOR

TWELFTH THOUSAND.

NEW-YORK:
ORANGE JUDD & CO., 41 PARK ROW

1866.

THE

GRAPE CULTURIST:

A TREATISE

ON THE

CULTIVATION OF THE NATIVE GRAPE

BY

ANDREW S. FULLER,

PRACTICAL HORTICULTURIST.

Price, $1 50.
SENT BY MAIL, PREPAID, ON RECEIPT OF THE PRICE.

NEW-YORK:
ORANGE JUDD & CO. 41 PARK ROW.

1866.

THE
ILLUSTRATED
STRAWBERRY CULTURIST:

CONTAINING THE

HISTORY, SEXUALITY, FIELD AND GARDEN CULTURE OF STRAWBERRIES, FORCING OR POT CULTURE, HOW TO GROW FROM SEED, HYBRIDIZING; RESULTS OF EXTENSIVE EXPERIMENTS WITH SEEDLINGS,

AND ALL OTHER INFORMATION NECESSARY TO ENABLE EVERYBODY TO RAISE THEIR OWN STRAWBERRIES; TOGETHER WITH A FULL DESCRIPTION OF NEW VARIETIES AND A LIST OF THE BEST OF THE OLD SORTS.

WITH RECEIPTS FOR

DIFFERENT MODES OF PRESERVING, COOKING, AND PREPARING STRAWBERRIES FOR THE TABLE.

FULLY ILLUSTRATED BY

New and Valuable Engravings.

By ANDREW S. FULLER,
Horticulturist.

TENTH THOUSAND.

Doubtless God could have made a better berry, but doubtless God never did.—IZAAC WALTON.

NEW-YORK:
ORANGE JUDD & CO., 41 PARK ROW.
—
1866.

DOMESTIC POULTRY:

BEING

A Practical Treatise

ON THE

PREFERABLE BREEDS OF FARM-YARD POULTRY,

THEIR HISTORY AND LEADING CHARACTERISTICS

WITH

COMPLETE INSTRUCTIONS FOR BREEDING AND FATTEN-
ING, AND PREPARING FOR EXHIBITION AT
POULTRY SHOWS, ETC., ETC.;

DERIVED FROM THE AUTHOR'S EXPERIENCE AND OBSERVATION

BY

SIMON M. SAUNDERS.

VERY FULLY ILLUSTRATED.

NEW-YORK:
ORANGE JUDD & CO., 41 PARK ROW.
1866.

www.ingramcontent.com/pod-product-compliance
Lightning Source LLC
Chambersburg PA
CBHW030313240426
43673CB00040B/1153